Industrial and Business Space Development

Implementation and urban renewal

Stuart Morley, Chris Marsh,
Angus McIntosh and Haris Martinos

London   New York
E. & F.N. SPON

First published in 1989 by
E. & F.N. Spon Ltd
11 New Fetter Lane, London EC4P 4EE

Published in the USA by
E. & F.N. Spon
29 West 35th Street, New York NY 10001

Typeset in Great Britain by
Leaper & Gard Ltd, Bristol

Printed in Great Britain by
St. Edmundsbury Press, Bury St. Edmunds, Suffolk

ISBN 0 419 14790 X

British Library Cataloguing in Publication Data

Industrial and business space development:
   implementaion and urban renewal.
   1. Great Britain. Real property. Investment
   I. Morley, Stuart
   332.63'24'0941

   ISBN 0 419 14790 X

Library of Congress Cataloging on Publication Data

Industrial and business space development:
   implementation and urban renewal/
   Stuart Morley . . . [et al.].
        p.   cm.
   Includes bibliographies and index.
   ISBN 0 419 14790 X
   1. Urban renewal—Great Britain.
   2. Real estate development—
   Great Britain.   3. Industrial promotion—
   Great Britain.
   I. Morley, Stuart.
   HT178.G7153   1989
   307.3'416'0941—dc20

Contents

About the authors   x

1   Introduction   1

PART ONE: The Investment and Development Market

2   The property market and the UK economy   9
   2.1   World trends   9
   2.2   Trends in Great Britain   9
   2.3   Employment trends   15
   2.4   The UK property market   16
   2.5   The evolution of the UK property market   17
   2.6   National and micro economic policies   20
   2.7   Taxation of property   22
   2.8   The property market since 1982   23
   References   24

3   Property investment: the rationale   25
   3.1   Investment property   25
   3.2   Investment criteria   26
   3.3   Prime property   28
   3.4   Secondary property   28
   3.5   Portfolio mix   28
   3.6   The location of investment property   29
   3.7   The advantages of direct property investment   31
   3.8   Disadvantages of property ownership   32
   3.9   Investment portfolio analysis   34

3.10 Conclusion 35
References 35

4   Finance for property development 36
4.1 Taxation and taxation allowances 37
4.2 Assisted areas 37
4.3 Private sector finance 38
4.4 Ratio analysis 39
4.5 Presentation of case 40
4.6 Interest payments 41
4.7 Fees 42
4.8 Leasing finance 42
4.9 Property companies 42
4.10 Investment trusts 44
4.11 Rights share issues 44
4.12 Debentures, mortgages and bonds 44
4.13 Joint ownership 45
4.14 Unitization 46
4.15 Financial institutions 47
4.16 Insurance companies 48
4.17 Pension funds 49
4.18 Owner occupiers (internal funding) 50
4.19 Conclusions 51
References 51

PART TWO: Property Components

5   Traditional light industrial development 55
5.1 Classification of planning uses 55
5.2 Low-tech and mid-tech buildings 57
5.3 Seedbed/nursery unit 57
5.4 Investors' attitudes 60
5.5 A standard traditional warehouse – industrial specification 62
References 66

6   Business use space 67
6.1 The classification of manufacturing buildings 67
6.2 Out of town offices 68
6.3 Defining science/research parks 69
6.4 Trade parks 71
6.5 Examples of business parks 71
6.6 Business use space – a guide to the specification 72
References 77

7   **Workspace and business centres**   79
     7.1   Introduction   79
     7.2   Types and characteristics   81
     7.3   Costs   88
     7.4   Special types   90
     7.5   The contribution of workspace schemes and business
           centres   92
     7.6   Conclusion   95
     References   96

PART THREE: Implementation at Local Level

8   **Local economic and employment development**   99
     8.1   Introduction   99
     8.2   'Mainstream' model   101
     8.3   'Radical–interventionist' model   103
     8.4   Local economic development initiatives   104
     8.5   Trends in local economic and employment development   108
     8.6   Conclusion   109
     References   110

9   **Government incentives to industrial and business space
     development**   112
     9.1   Introduction   112
     9.2   Policy background   117
     9.3   Government grants and incentives, 1987/88   123
     9.4   Recent amendments to grants and incentives   141
     9.5   Conclusion   146
     References   147

10   **The role of development control in implementing industrial and
     business space developments**   149
     10.1   Introduction   149
     10.2   The relaxation of development control   150
     10.3   Simplified planning zones   152
     10.4   The Town and Country Planning (Use Classes)
            Order, 1987   156
     10.5   Negotiated agreements   159
     10.6   Planning conditions and planning agreements   168
     10.7   Conclusion   172
     References   173

11   **Implementation by local authorities through land
     ownership**   175

11.1	Introduction	175
11.2	The involvement of local authorities in industrial development	175
11.3	Alternative methods of implementation	191
11.4	Methods and procedure of implementation	199
	References	204

12 Local authority financial and legislative powers — 207
12.1	Introduction	207
12.2	Capital expenditure and finance	207
12.3	Central government control prior to 1981	208
12.4	Central government control post-1981	212
12.5	Changes to the system of control of capital expenditure	221
12.6	Legal powers to acquire, service, develop and dispose of land and to provide assistance to firms	226
	References	234

PART FOUR: Development Appraisal

13 Private sector land ownership — 239
13.1	Introduction	239
13.2	Market research and the development industry	239
13.3	Demand assessment	241
13.4	Assessment of supply	242
13.5	Specific location and site appraisal	243
13.6	Financial appraisal – the basic approach	244
13.7	Residual valuation	246
13.8	Capital profit, development yield and rent cover	258
13.9	Refinements to the residual valuation	261
13.10	Cashflow approach	261
13.11	Sensitivity analysis	267
13.12	Scenarios and forecasting	269
13.13	Conclusion	272
	References	272

14 Public sector land ownership — 273
14.1	Introduction	273
14.2	Four slice and side by side partnership agreements — a comparison	274
14.3	Explanation of the ground rent calculation	277
14.4	Participation clauses and four slice agreements	280
14.5	Ground rents and premiums	285
14.6	Two slice and lease/leaseback agreements	287
14.7	Two slice arrangements	288

14.8   Lease and leaseback agreements   289
14.9   Direct development by local authorities   293
14.10 Conclusion   296
References   297

Index   299

About the authors

Stuart Morley MA BSc ARICS DipTP is Head of the Estate Management courses at the Polytechnic of Central London. Practical experience includes residential and investment agency, valuation, development and urban economic consultancy to the private and public sectors. He is a Principal Lecturer in property investment and development, valuation and appraisal. He is a contributor to various books on development and investment appraisal, office development and the future of the property market with particular research interests in public/private sector development partnership agreements, local economic development, investment and development appraisal.

Chris Marsh FRICS MRTPI DipTP DipCP is Course Leader of the Urban Estate Management degree at PCL. Practical experience includes development work with a group of housing associations, local authority planner in development plan and development control, private practice and consultancy. He is a Principal Lecturer in Planning, and has particular interests in inner city initiatives and the negotiation of planning gain. He is currently researching planning gain and advises local authorities involved in planning gain negotiations in a Planning Gain Consultants.

Angus McIntosh BA(Econ) MPhil ARICS is head of Healey & Baker's Research Department providing general economic background material for the firm's clients and partners, as well as specific property research and property performance analysis. Previously he was the assistant Investment Surveyor for Provident Mutual Life Assurance Association and Senior Lecturer at the School of Surveying, Kingston Polytechnic, and Principal Lecturer in the Department of Surveying, Portsmouth Polytechnic. He is the joint author of a major textbook on property investment (published by Macmillans).

Haris Martinos DipEng(Arch) DipTP MRTPI was Head of Planning Economic Development, London Borough of Hammersmith and Fulham, from 1976–1986 where he managed a wide range of urban regeneration projects. He has now formed his own consultancy, IDP, and is directing a major EEC programme on local employment development and carries out consultancy work for the OECD, the UK Government's Inner Cities Initiative and other clients on enterprise, employment and regional development projects.

1

Introduction

The last decade has witnessed unprecedented change in the basic character-istics of the economy of the United Kingdom and in the manufacturing sector in particular, resulting in historically high levels of unemployment. Not so long ago in 1971, Edward Heath's Conservative administration considered that a national unemployment total approaching one million (4% of the workforce) was politically unacceptable and initiated its famous volte-face, gambling that a major commitment to encouraging economic growth by demand management would resolve the issue. However, at the end of 1988, although unemployment had fallen from its peak of over 3 million in 1986, it still stood above 2 million, or 7.5% of the workforce, despite con-tinuous strong economic growth for the previous seven years. This figure is in fact deceptively low, in part due to short-term training programmes and in part to statistical changes made by the Government. Interestingly, the Department of Employment recently recalculated previous levels of un-employment and the latest revised basis now shows that unemployment in the early 1970s did not exceed 0.8 million while in 1979 when the Conser-vatives returned to power, unemployment totalled 1.11 million (4.1% of the workforce), one third of a million less than the previously published figures.

In many inner city areas and less favoured parts of the country, levels of unemployment are significantly higher than the national average and reflect the enormous problems of environmental dereliction, decaying infrastruc-ture and inadequate buildings, which urgently need to be overcome before prosperity can return to these areas.

The deep economic recession which initially followed the Conservative's election victory in 1979 adversely affected the traditional industrial sector more than any other and despite the years of economic growth since 1982, manufacturing industrial production only returned to its 1979 level in 1987.

The recession caused the total collapse of some traditional industries and substantial rationalization and restructuring of others, although the high 'shakeout' of employees that resulted, did encourage some individuals to utilize redundancy payments as investment capital and form new small businesses, an approach which was actively advocated by central government as part of their drive to promote an 'enterprize culture'.

Innovation was highlighted and the restructuring of the economy, coupled with the growth of information technology, led to an increase of knowledge-based industries. Such developments not only contrasted with traditional industrial investment by demanding a new and different type of accommodation, but their locational and environmental requirements were also at variance. Consequently, within and between the regions, a strong spatial discrimination in the economic welfare of particular localities has re-emerged as a major issue.

At the same time as innovation, growth and change have taken place in certain regions (primarily in the south of the country), many provincial areas, in contrast, have declined further, often with concentrated black spots in the inner urban areas. Inevitably, economic imbalances and opportunities encourage migration by the skilled and mobile elements within the labour market, a trend which of course exaggerates the distinctions and further limits the outlook for depressed areas. The so-called north–south divide therefore, together with the inner city problem have rapidly become a focus for profes- · sional, public and political concern, especially to local government increasingly involved in economic development and, arguably, rather belatedly, to central government.

The outcome of that anxiety has been most obviously manifest firstly, in efforts to restrict growth in the 'overheated' south-east, where suitable development land is in short supply and secondly, by a bewildering range of national and local initiatives, aimed at facilitating economic recovery in depressed areas.

From an industrial and business space perspective, the impact of economic change on the market has been particularly profound and contrasting. On the one hand, there has been strong demand in the south for new development to cater for high technology industries, often in a business or science park format and invariably located outside urban areas to benefit from good accessibility and environmental quality. In practice, however, such pressures have increasingly come into conflict with restrictive development plan requirements, while limited land supply has been further constrained when appropriate sites have been lost on appeal to higher earning uses. On the other hand, in the inner urban areas and less favoured regions, where the impact of change necessitates a substantial restructuring of the local economic base and urban renewal on a massive scale, demand for industrial and business space units has been relatively weak. Such circumstances clearly require a wide ranging economic development strategy. Property development, albeit subsidized where necessary, is an important contributory element to that approach.

Overall, the recent improvement in the national economy has promoted a gradual increase in demand for industrial property, a reduction in surplus space and a consequential rise in rents, especially since 1987. This in turn, has provided the necessary stimulus for private sector development activity and traditional industrial development (or sheds) is now once again occurring in some areas (market forces permitting), after a virtual absence for much of the last decade. Nevertheless, there are still areas of high unemployment and poor demand, which are unable to generate viable development schemes and, thus, private sector investment.

Throughout their three terms of office since 1979, the Conservative Government has sought ways to encourage the private sector to develop in areas of low demand with a variety of financial and fiscal incentives. At the same time, they have attempted through financial controls and the promotion of non-elected agencies, to reduce the role of local authorities in economic development, urban renewal and property development, most recently demonstrated in the Local Government (Prescribed Expenditure) (Amendment) Regulations, 1988. Local authorities, not surprisingly, due to the enormity of the problems many have faced and the political necessity for a high profile response, have fought these restrictions and devoted much time, effort and invention to overcoming or circumventing constraints. Despite this however, there has been a growing realization amongst local authorities, especially following Mrs Thatcher's third term election victory, that the scale, complexity and resource implications of local economic issues cannot be resolved in isolation and that some form of partnership with the private sector and other agencies is a practical necessity. In turn, that acknowledgement has prompted a new need-to-know attitude in terms of private sector investment and development criteria and thus, in order to promote industrial and business space developments, local authorities must have:

1. An up-to-date knowledge of development activity and emerging trends in order to understand and indeed anticipate market dynamics;
2. An appreciation of the rapidly changing requirements of industrial and business space users;
3. An immediate response to relevant Government policy reviews and their market repercussions, exemplified by the impact of the new Use Classes Order;
4. An intimate understanding of the developers assessment of financial viability and in particular, where the authority exercise a current or potential land ownership role, an awareness of the partnership options and financial arrangements evolving elsewhere, that could yield initial and future returns to the community;
5. A published policy statement of grants and incentives available and

regular monitoring of the effectiveness of previous partnership arrangements;

6. A sound knowledge of the financial and legal constraints on, and opportunities for, involvement and intervention in the property development
market and the role of development control in negotiating industrial and
business space inclusions in development proposals.

From a private sector perspective, more viable industrial and business
space development opportunities have emerged in the improved economic
climate of the late 1980s, reflecting strong demand in certain sectors, while a
variety of planning and financial incentives aimed at urban and regional
renewal are available to support otherwise marginal schemes. But, in order
to secure an effective and remunerative investment portfolio and development strategy, especially in the aftermath of recent sharp fluctuations in the
financial markets, the private sector requires detailed knowledge of:

1. The performance and growth of the national and local economies as a
framework for development decisions in response to user demand;
2. The sources and variety of development finance;
3. The content and location of viable projects and the most suitable
approach to development;
4. Appropriate appraisal and viability assessment techniques to guarantee
funding and reasonable profitability on freehold and leasehold sites;
5. The availability and range of grants and incentives;
6. The planning policy context, which determines local authority and
central government attitudes to development control, including the
most recent modifications such as revisions to the Use Classes and
General Development Orders.
7. The financial constraints on local authorities and their means of meeting
local needs and the opportunities this presents for developers to
negotiate lucrative deals.

This book, the first of its kind, addresses these and other industrial and
business space issues and is primarily intended as a practical guide to
implementation – essentially, how to get things done, whether by the
private sector, the public sector, or by a partnership between the two. In the
past, both sectors can be criticized for not understanding the motives and
requirements of the other. To be effective in today's competitive and more
deregulated system, such attitudes are, in our view, not only out of place but
a positive hindrance to economic rejuvenation and prosperity.

Aimed at both students and practitioners interested in industrial and
business space development, including estate managers, planners, architects, economists and geographers, the book is structured into four parts.

Part One presents a contextual overview of the property investment and
development market. Initially focusing on economic and employment trends

and their impact on the United Kingdom property market and, in particular, the demand for industrial and business space, evolving market conditions are analysed and indicative 'signposts' for the future identified. Subsequent chapters consider first, investment in property, the advantages and disadvantages of property ownership and the assessment criteria involved in property purchase and portfolio performance and management and, secondly, the options available to fund the development of industrial and business space schemes. The sources and characteristics of appropriate development finance are discussed, including the effect of government grants and tax incentives and the rapid expansion of funding initiatives such as debt and equity finance.

Part Two continues this theme, concentrating on the implementation of development proposals in three specialized sub-sectors, namely business use space, traditional light industrial/warehouse projects and workshop schemes. The evolution of such developments, their planning, layout and design characteristics and their funding are considered in detail, examples ranging from high technology science and business parks to inner city serviced, managed workshops and innovation centres.

Part Three in contrast examines public sector involvement in the industrial and business space development process. Building on the economic, social and political context that necessitated central and local government initiatives, the recent expansion of economic and employment development strategies are discussed. It then presents a critical analysis of the plethora of government incentives, aimed primarily at the public sector and including area-based programmes, such as enterprize zones and simplified planning zones and specific grants to individual projects, for example the City Grant and its predecessors, Urban Development Grant and Derelict Land Grant.

The role of local planning authority development control activity in implementing industrial and business space schemes is then explored in the light of the increasing prominence of government circulars, the changes to the Use Classes and General Development Orders and the growing importance of planning gain and legal agreements, as a means of securing industrial and business space elements in development proposals. The significance of other implementation vehicles available to local authorities is also emphasized, for instance where land ownership powers are exercised to provide serviced sites for the private sector or through development partnerships and, where possible, by direct development. Finally, the advantages and disadvantages of alternative methods of implementation and the necessary procedures and mechanisms involved are considered, within the important and highly topical context of capital expenditure controls and legal powers covering the acquisition, development and disposal of property.

Part Four concludes the book by providing a detailed examination of development appraisal techniques, employed by the private and public sectors to evaluate industrial and business space schemes, the process being illustrated by numerous numerical examples. The techniques discussed range from simple residual valuation methods producing capital site values, ground rental value and viability assessment, to more detailed equity sharing arrangements, cashflow appraisals and sensitivity/probability analyses.

As a policy objective at both national and local levels, the pursuit of economic rejuvenation and urban renewal commands general support, but the strategic options and subsequent means of policy implementation vary considerably, evolve continuously and inevitably reflect market expectations and political outlook.

In compiling this book, the authors intend to highlight the wide range of implementation methods available, as applied to industrial and business space developments, detail the procedural requirements and provide an insight into the skills needed to carry out realistic development appraisals. We recognize that market fluctuations and economic and political uncertainties may prompt further modifications in the future but trust that in focusing upon established and innovative approaches to implementation, we have made a meaningful contribution to understanding and appreciating the obstacles to, and opportunities for, successfully implementing industrial and business space developments in pursuit of urban renewal strategies.

Part One
THE INVESTMENT AND DEVELOPMENT MARKET

2

The property market
and the UK economy

2.1 World trends

There are a number of world events which are influencing changes in the
British economy and the property market is continually responding to these
trends. The purpose of this chapter is to relate national and international
events to changes in the property market in the UK.

Oil prices have had a more dramatic effect than any other commodity
price over recent decades. The increase in oil prices in 1973 and again in
1980 had very dramatic effects on the output of the countries involved with
the Organisation for Economic Co-operation and Development (OECD).
Industrial production dropped significantly during 1980 and continued to
fall to a low point in the final quarter of 1982. Since the beginning of 1983
production has slowly increased although the total output of the OECD
countries during 1986 increased slower than during the earlier period of this
growth.

Perhaps what is also relevant is that retail consumer prices (inflation)
have also shown a common pattern throughout the OECD. The increase of
more than 12% throughout the OECD in the first half of 1980 since when it
has been significantly lower. It fell very dramatically in the early part of 1986
when oil prices fell from $28 a barrel down to below $10 a barrel in a matter
of months. More recently retail consumer prices in all OECD countries have
either ceased falling or in some cases started to increase once more.

2.2 Trends in Great Britain

It is not surprising, with this background to world events, that the British
economy has also changed over this period. Within Great Britain these
events have been even more marked than in other OECD countries. When
oil prices trebled at the end of 1973, inflation increased from 7% to 22% over

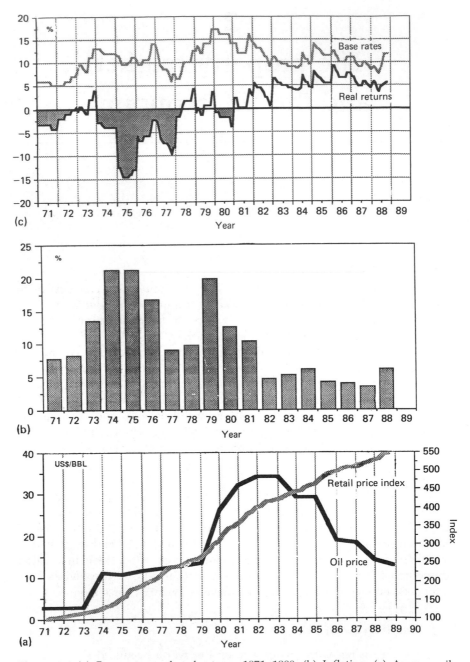

Figure 2.1 (a) Base rates and real returns 1971–1988; (b) Inflation; (c) Average oil price – US$/BBL (Saudi Arabian Light/Brent) and RPI.

Sources: CSO, BP Statistical Review of World Energy, June 1985; Petroleum Economist and Healey & Baker Research.

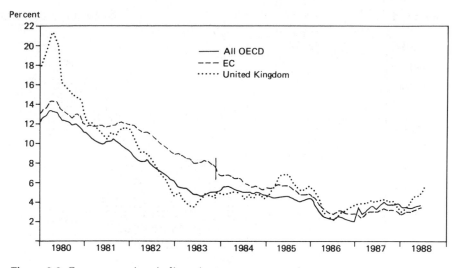

Figure 2.2 Consumer prices indices: increases over previous year.
Source: Employment Gazette, November 1988.

a period of 24 months. Although it fell to around 10% in 1977, it increased dramatically in 1979/80 to 20% again when oil prices doubled to 30 US dollars per barrel although oil prices were not the only cause of this jump in inflation. Oil prices fell for four years from a peak in 1982 and during this period UK inflation also fell to a low point of 2.5% at one point in 1986. In recent years oil prices have been far less influential on the world economy.

During the 1970s there were a number of years when real interest rates (i.e. monetary interest rates less inflation) were negative. This was not unique to Great Britain although the extent of the figures was greater than for many other OECD countries. In this period real assets, including many property investments, produced returns which were better than the low or negative returns experienced by other investments such as government stocks (gilts) or ordinary shares (equities).

In 1982 the financial market changed. Interest rates became positive with real returns of more than 5% being experienced in several years. During this period until 1987 many property investments performed badly as they were unable to meet this more demanding target. The equity market produced returns of 20% or more in a number of years whilst the demand for industrial and office property was falling. Thus the combined effect of attractive investment returns in the Stock Market and poor demand for property resulted in property investment yields increasing and a lack lustre performance from many investment portfolios.

For more than a decade the working population, as the graph shows, has been steadily increasing with the exception of a period between 1980 and 1983. However, it is noticeable that, although the employed labour force fell very dramatically between 1979 and 1982, since that date, in line with the

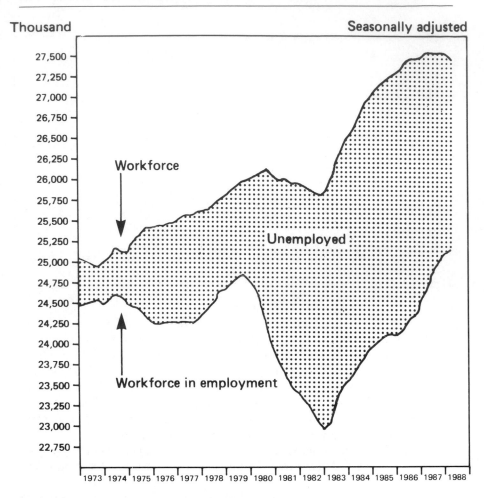

Figure 2.3 Workforce and workforce in employment: Great Britain.
Source: Employment Gazette, November 1988.

increase in world production, employment in Great Britain has been steadily increasing. The problem is that employed labour has still not caught up with the size of the working population, hence unemployment.

Partly as a result of the hike in oil prices and the down-turn in production in the 1980–82 period, unemployment in Great Britain rose very significantly. During 1987 unemployment started to fall from its peak in 1986. The dilemma is that the level of job vacancies still falls considerably short of the number of unemployed.

For the property market, the significant trend relates to the structural changes which have taken place in the British economy. The graph shows the steady decline of manufacturing employment in Great Britain particu-

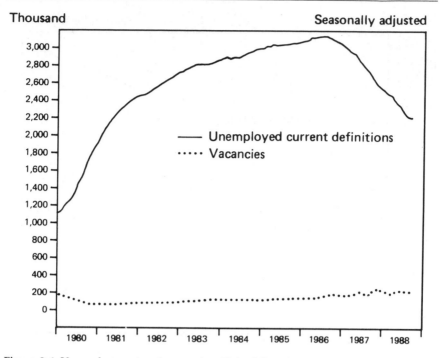

Figure 2.4 Unemployment and vacancies: United Kingdom.
Source: Employment Gazette, November 1988.

larly after 1979 although there is evidence that, in the last three years this decline in manufacturing employment has levelled off. Non-manufacturing employment also declined in the period between 1979 and 1983. However, as the graph indicates this is now steadily increasing. In fact it has returned to the pattern of increase which was witnessed throughout the 1970s.

There are three long-term factors which need to be taken into account in understanding the structural changes taking place in the British economy.

The first relates to the growing level of affluence. Although there have been years when the growth in spending has been low, in fact below the rate of inflation such as between 1979 and 1982, long-term spending continues to rise at between 2% and 5% faster than inflation. Part of this affluence is moving into the growing owner-occupied residential market but much of it goes into retail expenditure. It also means that we, as a nation, demand higher quality buildings: at home, when we go shopping and, importantly, at our place of work.

The second factor which is equally important is the growing mobility of the population. The number of cars on Britain's roads is increasing at between 2% and 3% every year. This is in common with many OECD countries although some countries, such as Japan, have seen far more dramatic increases in car ownership. But looking at Great Britain alone this

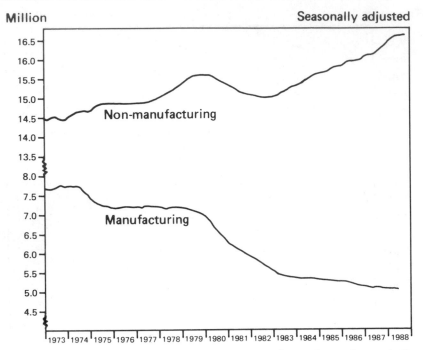

Figure 2.5 Manufacturing and non-manufacturing employees in employment: Great Britain.
Source: Employment Gazette, November 1988.

increase means that over the 30 years between 1970 and the year 2000 the number of cars on Britain's roads will have increased by 100%. All industrial and business property development has to take this factor into account in both its location and design.

The growth in air travel is even more dramatic and has recently been rising at around 5% per year. Mobility is changing our concepts as more people travel overseas on business or for holiday. But it is also changing the nature of the industrial and business property market as an increasing number of firms want to be located with easy access to an international airport and hence an international market.

The third factor which is influencing all our lives is the increasing amount of leisure time. For many people a working week of more than 40 hours is a thing of the past, flexitime is part of the work experience and most people now enjoy at least four weeks holiday per year. With more leisure time employees have the opportunity to enjoy the environment within which they live. There appears to be a preference to live in certain parts of the country rather than others hence making modern British industry highly locational conscious. A number of surveys have found that the locations they choose is where the workforce would prefer to live and enjoy its leisure hours. The southern half of Great Britain is generally preferred to the north.

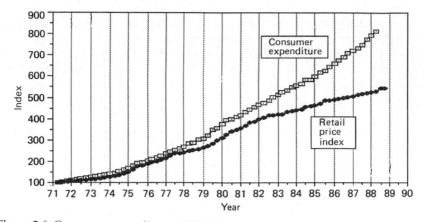

Figure 2.6 Consumer expenditure v RPI 1971–1988.
Source: Central Statistical Office.

2.3 Employment trends

As Table 2.1 shows the numbers employed in all industries in Great Britain increased by 3.7% between 1971 and 1980. However that level of employment *decreased* between 1980 and 1988 by 4.5%.

For the changes which are now taking place in the property market, the relevant factor is the nature of employment. Manufacturing industries shed 14% of their employment between 1971 and 1980 but during the last 8 years the employment decline has been much faster at 26%.

The converse has happened with service industries. During the 1970s service industries were increasing at a very rapid rate. In the years from 1971 to 1980 employment in this sector of the economy increased by 18.6%. What is significant is the employment in this sector, whilst it has continued to increase, between 1980 and 1988 the increase at only 8.2% was much slower than during the previous decade. The increase of 1.1 million employed in the service sector does not replace the 1.8 million jobs lost in the manufacturing sector since 1980.

Table 2.1 Employment trends in Great Britain: million persons

	1971	1980	1988	1971–80	1980–88
All industry	21.6	22.4	21.4	+3.7%	−4.5%
Manufacturing industry	7.9	6.8	5.0	−14.0%	−26.0%
Service industry	11.3	13.4	14.5	+18.6%	+8.2%

Source: Employment Gazette, HMSO.

The other trend which is important for the property market, is the growth of small businesses. Whilst the government does not keep any accurate statistics on the growth of small companies, from taxation returns they do know that the number of self-employed persons within Great Britain has been changing. Whilst in the years between 1971 and 1980 the number of self-employed decreased very slightly, since that date the number of self-employed has increased by a dramatic 45.4% (Table 2.2).

Also relevant is the continual drift south of the population with the East Midlands, East Anglia, South-East England and South-West England continuing to show population increases, a factor which is once again reflected in the industrial and business use property market.

Table 2.2 Employment trends in Great Britain: self employed: thousand persons

1971	1980	1988	1971–80	1980–88
1952.8	1950.2	2835.0	−0.13%	45.4%

Source: Employment Gazette, HMSO.

2.4 The UK property market

Over the last 20 years the property investment market in the United Kingdom has reputedly become the most sophisticated in the world. It did not develop in this way by design. No government introduced legislation with the specific aim of encouraging the type of property market which exists today. The property market evolved due to a number of influences and in response to a number of situations. Inflation, national economic policies, world trends, entrepreneurial skill, the poor performance of gilts and equities during the 1970s, town planning and taxation have all contributed in different ways to the growth of the present property investment market.

To state that the market developed in response to the high inflation in the 1970s is an insufficient explanation. All western countries based on a market economy experienced inflation to a greater or lesser extent over this period. Alternatively, to state that it was the growth of investing institutions as financial intermediaries which caused the property market to develop is also an inadequate explanation. All OECD countries have experienced a growth in pension systems and many have used the funded pension scheme as a way to meet this objective. Whilst France and Germany developed a pay-as-you-go pension system, the United States of America in particular developed a funded pension system although it did not develop the type of property investment market common in Great Britain.

2.5 The evolution of the UK property market

Fixed interest investments, such as government securities and mortgages, were in the past the traditional media through which life insurance companies would invest. The investment risk was relatively low and the return on investments, although not specifically high, was relatively secure.

Mortgages were granted for a period of between 10 and 40 years on a fixed interest basis. Throughout the life of the mortgage the interest rate and periodic repayments remained unchanged. In times of inflation the problem of this system was that the mortgagee received none of the increase in capital and/or rental value of the property which changed during the term of the mortgage.

Initially, to overcome this problem, institutions granted mortgages on variable rates of interest to developers; the interest payable was adjusted periodically and became related to the rental value of the property. However, institutions soon realized that they would have more control over their investment expenditure if, instead of granting a mortgage loan based on the collateral value of the property, they purchased the investment and granted a leaseback to the developer.

By the late 1960s the sale and leaseback method of funding property development had become common practice. The developer guaranteed to pay a leaseback rent to the institution depending on the nature of the investment. Initially, leaseback rent was based on fixed interest rates. However, as the rental income increased, the developers share of this rent increased over time. In other words, the institutions were often in no better position than if they had granted a fixed interest mortgage themselves even though they owned the freehold or long-leasehold interest of the building.

Institutional finance became popular with developers because it was outside the government monetary qualitative controls during this period. For occupiers, it was an attractive way to raise finance to meet liquidity requirements, to meet debts or for further expansion.

Various amendments to this simple sale and leaseback method to developers were therefore used. Sometimes the increase in rental value, over and above the initial rents when the building was first constructed, was shared in agreed proportions between the developer and the institution. Later, institutions insisted on receiving a fixed percentage, say 80%, of the property's rental income. It should be remembered that the developer was technically the head lessee who underlet the building to an occupying tenant.

Although the change from mortgage interest arrangements to institutions owning an equity interest in the property enabled the institution to benefit from inflation in the value of rents over time, the security of income was suspect. If the developer went into liquidation or no tenant was found for the building, an institution could find that it had invested many millions of pounds in a property which was producing little, if any, return. These problems reached a peak in 1974 at the time of the property market crash.

In many ways, the property boom and its subsequent collapse encapsulated the principle influences which made financial institutions into the dominant landlords they are today.

The boom was partly created by the high level of demand, relative to a fixed supply, for property during the late 1960s and also the relaxation of monetary supply and development controls in the early 1970s as the new government made a dash for growth before the UK became a full member of the European Economic Community. In those inflationary times, property seemed a very good investment medium and in many ways looked a better bet than fixed interest stocks or company equities.

It was at this time that the institutional lease became more sophisticated. The modern five year rent review pattern became widely accepted by the property market. At the same time, the system of Office Development Permits (ODP) and Industrial Development Certificates (IDC) existed in south-east England. The famous George Brown ban on office development in central London in 1964 was an example of this restriction on development. Developers therefore sometimes commenced development programmes outside south-east England. Towns such as Birmingham, Bristol, Derby, Leicester, Manchester, Nottingham, Portsmouth and Southampton all witnessed office construction booms. Even in the early 1980s, ten years after much of this construction, many of the buildings constructed during this boom period remained unlet. Institutions and banks provided the sale and leaseback finance to the property company. The finance was provided on the assumption that rental values would continue to rise and the speculatively built buildings would readily let when completed.

During 1973, a series of events caused the market to collapse. In an effort to control the economy and to prevent a balance of payments crisis, the Bank of England minimum lending rate increased from 4.5% in early 1972 to 13% by the end of 1973. Such a tight monetary policy was at the expense of property companies although investing institutions were far less vulnerable.

As a result of inflation building costs also rose dramatically. Together with the cost of short-term finance, the total cost of new buildings increased very much faster than rents. But at the same time the level of demand for buildings under construction fell very dramatically. The oil crisis in early 1974 caused crude oil prices to rise by 300%. But monetary policy had also dampened economic activity, reducing the overall demand for property in the UK economy.

It was two pieces of government legislation which finally caused the collapse. A rent freeze was introduced as a way of controlling prices. Acting under political pressure to tax property companies, Development Gains Tax legislation was also introduced in December 1973 but, by this time, property values in general were already falling. This legislation was more like shutting the stable door after the horse had bolted and, in the event, the legislation caused the stable to collapse! The loss of confidence was so dramatic that the Bank of England launched its so called lifeboat operations to save the banks suffering as a result of property companies going into liquidation.

It was the institutions who became even more involved in property as a result of those dramatic events. During the boom period, many institutions purchased investment properties for the first time in an attempt not to miss out on what appeared to be a lucrative medium for investment. In 1974, it was the large institutions themselves which saved the banking system by lending the banks money and by purchasing property previously owned by property companies which by then no longer existed. In other words, the institutions were very much part of the lifeboat operation.

The property crash was caused by a property boom followed by falling demand, a rent freeze, harsh monetary policies and legislation to tax property development. The result of these events was to make the UK investing institutions greater owners of commercial and industrial property than in any other country in the world. Hence they became increasingly influential in the shape of the UK property market.

It was not until the early 1980s, when demand for office and, in particular, industrial space fell dramatically as a result of the recession, that the influence of institutions in the market began to wane. Between 1982 and 1988 the banking sector came back into the market and bank loans outstanding to property companies increased from less than £3 billion to more than £18 billion. But by the end of 1987 there was clear evidence that institutions

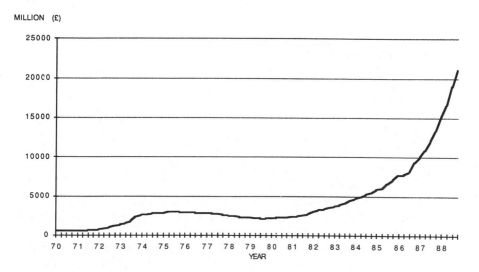

Figure 2.7 Bank lending to property companies.
Source: Bank of England.

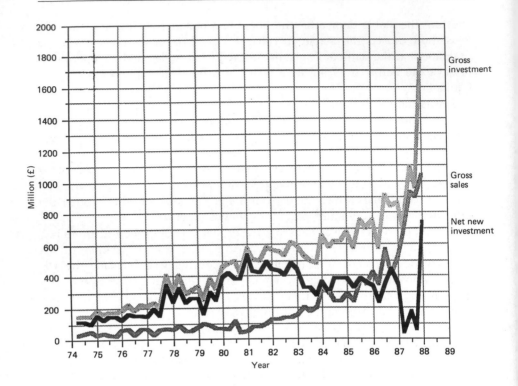

Figure 2.8 Insurance companies and pension funds; quarterly property investment transactions.
Source: Central Statistical Office.

were once again increasing their net investment in the property sector of the investment market.

2.6 National and micro economic policies

The shape of the UK property market has also been influenced by other government legislation. When Government passes acts of parliament to achieve a stated objective, they indirectly and often unknowingly cause other changes although the changes may not be immediate but may take years to reach fruition.

One of the features of the Town and Country Planning Act 1947 was to impose a development charge of 100% on the increase in value of land resulting from new development. Although this piece of legislation was repealed, there have been several attempts since 1947 to impose a similar development charge of tax on development of land. All such enactments relating solely to land have now been repealed.

In a market economy, for production to take place, a surplus of profit must be produced. In other words, the sale price realized by a product must be greater than the cost of producing that product. In a planned economy, where production is often dictated by hegemony, market forces are normally unimportant as a way of determining people's wishes.

For a variety of reasons land has been singled out for taxation on a number of occasions but it was the 1947 Act which had the most dramatic effect on the market. This was coupled with the need, due to the shortage of materials in the postwar period, for building licences before development could take place. The problem was that in the immediate years after the war the economy expanded faster than the supply of buildings. Hence there was pressure on the government to repeal the system of building licences which eventually disappeared in November 1954.

The economy continued to expand so fast that, within 12 months of 1954, a stringent and lengthy credit squeeze was started with bank rates rising from 3% in mid-1954 to 7% in September 1957. In other words the cost of short finance necessary for building construction had more than doubled. Controls were also imposed on banks but then, as now, banks were only one type of finance.

Insurance companies as previously explained were largely outside the control of the Government's monetary policy. Publicly quoted property companies therefore turned their attention to insurance companies as a source of funds. As previously discussed, insurance companies therefore became increasingly involved in providing finance for property develop-ment. Indirectly and unknowingly the government of the day had encou-raged life insurance companies to become involved with the property market.

The credit squeezes of the 1950s and subsequent times have had other effects on the property market. At times of credit squeeze the manufacturing and trading companies have often experienced liquidity problems; the income of the company from the sale of products has been insufficient to meet outgoings including short-term borrowing from the banks. Companies have often found themselves in a highly geared situation; the ratio of debt capital, often bank loans, becoming disproportionately large in relation to equity capital.

One way that manufacturing and trading companies have found for improving their liquidity problems and reducing the interest payments on debt capital has been to realize the value of some of their assets. The disposal of freehold interest in return for leaseback has been one way, in the past, of realizing capital to reduce borrowing. Over the last 20 years both large and small manufacturing trading companies have sometimes resorted to this financial tactic to preserve themselves. Once again government monetary policies have indirectly and unknowingly often encouraged industrial and commercial companies to hand over the freehold of their property to life insurance companies.

Property played an increasingly important part over the 1960s and 1970s in many takeover battles. The concept of asset stripping became prevalent in cases where the land value of trading companies was not accurately reflected in their account.

Particularly during the 1950s and 1960s companies often reflected the value of their assets at their original cost. Also the trading company was paying only a small dividend so the market value of the companies shares could well be relatively low. The potential gain from purchasing the trading company as a going concern at a low market value and selling off the property assets and realizing the full value was often enormous. Hence, on the one hand trading companies were often forced into the situation of selling off their freehold and leasing back from investing institutions, and on the other hand investing institutions and property companies often indulged in the process of asset stripping so that the full value of property assets could be realized.

2.7 Taxation of property

A number of pieces of taxation legislation over the decades have also changed the position of investing institutions in the property market.

The Finance Act of 1921 gave certain tax advantages to pension funds and life insurance companies running pension funds or an annuity business. As a result of the act not only can contributors offset contributions to a fund against Income Tax, but the earnings of the fund, received from investments also accumulate free of tax. The privileged tax position of pension funds and life insurance companies involved with this sort of business still remains today. However, in 1984, the government removed the benefit which could accrue to tax payers when making contributions to their own life insurance policies.

It was the change to Corporation Tax in the Finance Act of 1965 which for a time dramatically changed tax status of taxable companies. The 1965 Act introduced the concept of corporation tax. This taxation had two elements; Corporation Tax on income which rose to 52% by 1984 but fell back again to 35% by 1986 and Capital Gains Tax. This was a tax on the capital gain after 1965 between the original cost of purchase of an asset and the eventual sale price. In 1982 this onerous burden, which in effect became a tax on inflationary gain rather than real value changes, was amended. The concept of indexation to the level of inflation was introduced such that Capital Gains Tax in recent years has only been payable on the surplus gain over and above inflation between the date of purchase and the date of sale. Since 1988, all gains before 1982 have been removed and are no longer liable to CGT.

The important aspect of this taxation is that pension funds and pension fund and annuity business of life insurance companies continue to be exempt from such taxes. In other words, financial institutions continue to be in a favourable tax position as far as the ownership of property is concerned although it is significantly less favourable than it was at the beginning of the 1980s. The imposition of Value Added Tax on all building works will also go against the interests of institutions, particularly those involved with property development. As they are exempt from VAT they will find it difficult to offset this tax burden.

2.8 The property market since 1982

1982 has become a watershed in the UK property market. As a result of the increase in oil prices in 1980 and the downturn in world trade, demand for property fell away quite dramatically. The recession in the economy, and the property market in particular, resulted in many institutions reassessing their investment attitudes to property.

Since 1982 the tax regime, as previously mentioned, has been far more favourable to property companies and owner occupation. During the early 1980s the investment market has generally witnessed a dramatic improvement in investment performance with equities often showing returns in excess of 20% per annum until October 1987, whilst property performance was often below 10% per annum. In 1988 property investment returns exceeded 20% for most portfolios.

It was the new entrepreneurial property companies who were able to benefit most from the changes in the property market as they had not had to suffer the inertia of previous investment decisions. By the mid-1980s the market had changed quite dramatically with the financial institutions not dominating the property investment market to the degree of the 1970s. An increasing number of industrial and office buildings have become owner occupied ranging from City of London office buildings being owned by those who trade from within them to industrial buildings often being purchased back from life insurance companies and pension funds who had acquired them a decade earlier.

The more astute property investment companies have also returned to the investment scene, a position from which many had retreated in the 1970s. With the changes taking place in the property market, the banks have also returned to the market after their unfortunate experiences in the period 1973–1974. Hence, in the second half of the 1980s, there is a greater variety of finance and investment opportunities than a decade earlier. Investing institutions are still in the market but property companies are far more active as developers, traders in property and investors. The merchant banks are also providing far more finance for property development than for more than a decade.

By 1988 it was clear that, not only in the central London office market, but throughout the country, the property market was into a new stage in the market cycle. The slow increase in employment since the low point in 1982 had eventually absorbed the many surplus office and industrial buildings to such an extent that rental values started to rise quite dramatically encouraging developers once again to consider the advantages of office business space and industrial investment throughout the country.

At the same time many investing institutions who had disinvested from property in the early 1980s, rearranged their existing portfolios or only acquired retail property, were once again endeavouring to increase their exposure to the retail, office and industrial property investment markets.

References

Cadman, D. and Cataleno, A. (1983) *Property Development in the UK – Evolution and Change*, College of Estate Management Property Development Library, Reading.

Darlow, C. (ed.) (1983) *Valuation and Investment Appraisal*, Estates Gazette, London.

Darlow, C. (ed.) (1986) *The London Property Market in AD 2000*, E. & F.N. Spon, London.

Healey & Baker (1987) *Predicting the Property Market*, Healey & Baker.

Rose, J. (1985) *The Dynamics of Urban Property Development*, E. & F.N. Spon, London.

3
Property investment: the rationale

The most important criterion influencing the purchase of any property for investment purposes is location. This is the most crucial factor determining present and future rental and capital values. This chapter provides a general guide to the principal locational and design criteria which influence property investment decisions. It is an important area to understand as many industrial and business space developments are sold as investments to financial institutions and property companies.

3.1 Investment property

All investment is concerned with forfeiting the use of resources, normally financial, at a moment in time in the hope that there will be a return on that investment in the future. The return may be in terms of capital and/or income. In this respect property is like any other investment; an investment property is expected to generate rental income and increase in value as a capital asset. However, every investment involves risk and generally speaking the level of risk is commensurate with the expected level of return. The higher the risk the greater the level of return.

There is a subtle difference as far as the property market is concerned, between long term and short term. Short-term investment involves spending money on property or property development where the risks are relatively high yet the potential financial gain is also high. This may be because, for instance, the property is unlet at the date of the purchasing decision or it does not have planning permission for the intended use. Long-term investment, on the other hand, is concened with perhaps more modest financial rewards where the risks are relatively low. Sometimes, unknowingly, an investor inadvertently becomes a trader/short-term investor due to an error in his investment decision. Generally, though, an investor is seeking a secure investment where there is a relatively secure return expected from the investment.

Investing institutions fall into this latter category. Unlike property companies, they do not normally seek short-term property developments. Instead they generally prefer to purchase property which will form a sound financial investment over a number of years. This is not to say that some institutions do not indulge in short-term property investment decisions. Property performance league tables published by a number of organizations in recent years have put pressure on institutions to seek property invest- ment situations where there is an early financial gain. A property where the rent review is imminent is one such potential situation.

Investment institutions do however undertake development but rarely purchase sites which do not have the benefit of planning permission and where the chances of letting the eventually constructed building are remote. Instead they either purchase commercial and industrial buildings which are built and let, or they become involved in well-secured development situations. Such a situation might involve a site with the benefit of planning permission where the proposed building is pre-let to the eventual occupier. In other words a potential tenant has already signed a legal document agreeing to take up a lease and pay rent when the proposed building is complete. Even in this situation there are risks. The building may never be satisfactorily completed or the potential tenant may go into liquidation before the building is complete. The so called 'covenant' of the tenant is crucial to the investment decision. The rental income to the investor is paid by the tenant and, therefore, to reduce the investment risk, institutions have a strong preference for publicly quoted companies or public authorities as tenants.

Property investment is essentially concerned with investing in property which will produce a secure rental income over a number of years and which may also gain in capital value.

Investors concentrate on investing in offices, shops, warehouses and industrial property. In addition, many investors have purchased property overseas.

3.2 Investment criteria

The most important criterion is that the property purchased will let and generate a rental income. In this respect investment surveyors, when considering a property to purchase or to value, look at the property in relation to a number of headings:

(a) Location

This is the most important factor. Even if the present tenant goes into liquidation a well located property should re-let relatively quickly. For this reason the property investment market often states that the three things of importance when purchasing an investment property are 'location, location and location'. Even a badly designed building will let if it is in a desirable location.

(b) Design

This is the next most important consideration and has become increasingly important and must be examined in relation to the property's intended use. The property's appearance, the functional space it provides and the cost involved in using the property all have to be considered.

(c) Tenure

The nature of the tenure and the legal restrictions which may be imposed on the land are important. In this respect freehold property is preferred to leasehold property. However there are a variety of long leasehold arrangements which can be purchased by investors; the exact nature of the leasehold can be important in determining the nature of the rental income. In practice landlords and tenants (long leasehold investors) rarely have the same ambitions and it is therefore difficult to cater for all eventualities. When disposals, refurbishment or redevelopment are considered problems sometimes arise which encourage investors to favour freehold investments.

(d) Tenancy

The actual tenant and the lease document (which makes that tenant a contractual party to the investment) are both important. As previously mentioned, whether the tenant is a private individual, small company or publicly quoted corporation can influence the level of investment risk. The lease itself dictates the nature of the rental income. A badly worded lease can dramatically influence the rent received at rent review and hence the rental expectation from the investment.

(e) Occupational lease

The occupational lease contractually links the investors asset to a stream of income (rents) from the tenants. In the UK a standard lease is for 25 years with the provision that every five years the rent may be reviewed to the open market rental value at that time. Most leases now specify a procedure for settling the rent at the rent review date if the two parties are unable to agree the new figure. Normally the President of the Royal Institution of Chartered Surveyors will be called upon to appoint either an expert surveyor or an arbitrator to settle the rent.

The income received by the investor is net or clear rent. The tenant is responsible for fully repairing and maintaining the property and is also responsible for insuring the structure. Where the building is multi-let as in the case of a shopping centre or office building with several tenants, the insurance, management and repair may be undertaken by the investor for all the tenants. He then charges a service charge rent in addition to the rent paid for the use of the premises.

3.3 Prime property

The concept of 'prime' property is an integral part of this method of analysis involving the above categories. Inevitably there is an element of judgement involved in defining whether a property is 'prime'. Simply stated, a prime property is one which is located in the best position, is new and well designed, is of a lot size (value) which is popular with investors, is freehold with no restrictive covenants and is let to a sound tenant on a modern well drafted lease.

A number of organizations involved in the property investment market regularly publish their assessments of the prime investment yields applicable to each of the three categories of property, namely office, shop and industrial. A yield in this context relates to the annual return which may be expected from a rack-rented property expressed as a percentage. It is important to note that not only are these assessments of prime yields subjective but they are based upon a knowledge of the market. The concept of prime only relates to a very small section of the investment property market; perhaps only 1 or 2% of investment property can be considered prime in every sense.

3.4 Secondary property

Many investment properties fall into the category of 'secondary'. Again such a definition is subjective, but, given that there must be by definition a dearth of prime property, the majority of investment property is secondary, although some property is more secondary than others. In other words prime yields are not applicable when assessing a secondary property's capital value. Although many property companies specialize in secondary property, some institutions also invest a proportion or, in a few cases, the whole of their portfolio in such locations.

3.5 Portfolio mix

An examination of investment portfolios reveals that there is a significant variation in the proportion of funds allocated to the principal investment categories. For instance, some portfolios specialize in retail property whilst others have a bias towards industrial property. Having the correct mix of property is crucial, particularly for unitized market-orientated funds. An unquoted occupational pension fund has less pressure on short-term analysis and can, therefore, invest in property which may be expected to show a good long-term return yet in the short term may not perform particularly well.

An example of the mix which might be found in a property investment portfolio is as follows:

Type of property	Percentage of portfolio
Shops	50–70
Offices	15–25
Industrial and warehousing	15–25

Many existing portfolios have a much higher content of office and industrial property. It is difficult and perhaps misleading to consider a property portfolio simply in terms of the above classifications. Some funds deliberately buy short-term leasehold investments or high yielding stock where the strength of the lessee and the high investment yield is more important than the building's location, its design or expectation of rental growth.

3.6 The location of investment property

When considering an investment, the location, as already stated, is still the most important factor for most properties. The property will be considered in relation to the country it is in, the region within that country, the town or city within that region, the area within that town or city, and finally the road and position within that road. Understanding financial geography is critical to property investment.

An understanding of road, rail and air communications is vital in the consideration of location. Understanding such things as the influence of car parks and motorways on property values is important particularly when there are plans in hand to change the pattern of roads and car parking. The M25 London Orbital Motorway has had considerable influence on property values and business location.

Rail transport is most important when considering office investments. It is one of the ironies of modern cities that down-town or central business district land use is dependent on an efficient public transport system. Normally this includes a rail network. Whilst land values increase and businesses may prosper in the central area, private market forces are rarely sufficiently strong to make the transport system profitable.

The growth of air travel has significantly influenced the pattern of land use in the last two decades. At the heart of the Golden Triangle on the west side of London is Heathrow Airport. 'Golden' is applied because land and property values have increased significantly in recent years within this particular area and High Wycombe, Guildford and Hammersmith form the apices of this triangle. The M4 motorway from London to the River Severn bridge represents the Western Corridor or Silicon Valley due to the growth of electronics-orientated industry along this route. The influence of Heathrow Airport has also significantly aided this development and growth of property values.

The art of good property investment is to assess the property situation where the investment risks are minimized yet the potential for growth, in capital and rental value, is maximized. Although it involves scientific methodology it may be considered an art as it requires appreciation of the location and other factors in relation to the price of the property being considered. The price of any property is related to the rental value. Thus the art of good investment involves anticipating those locations and properties which are going to show rental growth which is above average for similar properties and hopefully above the Retail Prices Index (RPI).

When considering location it is important to note that the polarization effect caused by economic behaviour is often reinforced by government enactments. Town planning restrictions may reinforce the natural shortage of land for development in these sorts of areas by either preventing development or only permitting a particular kind of land use on certain conditions. The following is an example of one such policy:

During the plan period, office development proposals will be restricted to appropriate locations as defined in (the) policy . . . and will be considered within the overall context provided by the following principal factors which will be formally monitored annually by the local planning authority:

(i)   town centre and site environment considerations
(ii)   the capacity of the transport system
(iii)   the availability of car parking spaces
(iv)   the availability of labour
(v)   the demand for office accommodation
(vi)   the effects of new office technology
(vii)   the achievement of planning advantages in accordance with appropriate council policies and priorities
(viii) vacant office floorspace and outstanding unimplemented planning permissions for office development

Source Kingston Town Centre District Plan – written statement, July 1982.

Other planning authorities insist that a named occupier is provided with the planning application or that the building, when eventually built, should only be used by a company already located in the borough or county.

The effect of such policies is to reinforce the market forces and make property investment in such locations even less risky. With a restricted level of supply there is more likely to be rental growth.

Understanding the forces of supply and demand for particular property in a particular location is important for an appreciation of property invest-ment. Having analysed the other buildings which are similar in a particular locality, the recent trends in the letting market and the rental values paid for similar accommodation, a property investor has to apply considerable subjective skill in valuing a property and/or deciding whether to purchase a property for investment purposes. The following sets out the advantages of investing in property.

3.7 The advantages of direct property investment

1. The investor has more control over the properties purchased.
2. When large sums of money need to be invested, fewer acquisitions need be made as each building may cost several million pounds. In other words, large sums can be invested in a small number of individual units.
3. Property values are generally less volatile than shares and can usually be expected to rise but not necessarily in line with inflation.
4. When the stock market is depressed it is still normally possible to sell property at a reasonable value.
5. Property companies are often highly geared making their return less secure than other shares or direct property investment.
6. The dividend on equities is paid half-yearly in arrears yet most commercial property leases nowadays require the tenant to pay rent quarterly in advance.
7. Dividends on equities are taxed at source: the investor only receives dividend income net of tax, and tax-exempt funds must subsequently recover the tax. Property rent is paid gross before any deductions for taxation are made. For gross investment funds, such as pension funds which are exempt from tax, this makes a significant difference to the return on the investment.
8. Rent is paid by a tenant even if that tenant, as a company, is making a loss.
9. Rent is the first call on a tenant's income and is paid even before interest on debentures or bank loans. For this reason, rent is a more secure form of income for an investor.
10. Even if the tenant goes into liquidation, the investor still has an asset; the property itself. Following liquidation, a company's equities often only have a minimal, if any, value.
11. Most modern commercial property leases provide for the rent to be reviewed in an upward direction every five or possibly three years. The investor therefore, has a way of making sure his income is reviewed to

currently prevailing market levels at regular intervals, regardless of the tenant's profitability. Such a system reduces the investor's risk in times of inflation. Because the tenant pays rent and does not need to raise mortgage finance to purchase his property, he also can reduce his risk. The financial gearing of the company is improved as the company's debt capital in relation to its equity capital is improved. The interest on mortgage finance is particularly susceptible to variation, often as a result of government monetary policy. Such interest rate changes increase the risk of using mortgage finance to own property.

12. A vast amount of commercial property is occupied by private companies, or other organizations, who do not raise finance by issuing shares. It is therefore not possible to purchase equities in these companies. Through the medium of property investment, it is possible to benefit financially from the growth of certain sectors of the economy. The expansion of building societies and mutual life assurance companies in the last 20 years is an example. By purchasing certain property favoured by these expanding organizations it is possible for an investor to benefit financially from the demand for property created by them.

     The expansion of government departments is another example of the same phenomenon. During the 1960s, the government was often referred to as the 'developer's friend'. With the growth of the Civil Services and the mushrooming of salaried employment, the government demand for offices increased. The developers and the investing institutions obliged and discovered a lucrative medium for investing their funds. Whilst government departments often do not maintain buildings to the highest standard, the rent they pay is secure. Unlike a private or publicly quoted company, governments do not go bankrupt. Whilst the interest paid on gilt-edged securities remains unchanged throughout the life of the stock, the rent government departments pay for office accommodation is generally subject to reviews at five-yearly intervals.

3.8 Disadvantages of property ownership

Commercial property ownership clearly has certain advantages as an investment medium as indicated above when compared with equities, including the equities of property companies. However, there are also some potential problem areas which need to be noted. The following sets out some of these, which are in no significant order of priority.

1. Property investment has a very low liquidity. It is often time consuming and expensive in professional fees and stamp duty to purchase a building. Even at times of significant inflation, property values only increase gradually. It may take many months to sell a building and selling is also expensive in terms of professional fees. The problem is best explained with a simple example.

Imagine a property cost £1 million to purchase. The fees and stamp duty cost the purchaser approximately 2.75%. The total cost is therefore £1027500. The property has got to increase in value by nearly 5% before the purchaser can recover all his capital costs of purchase and sale. Suppose that he sells the same property for £1050000, representing an increase in value of 5%. As vendor he incurs sale costs of about 2%, made up of surveyor's and legal fees. In other words he receives only the net figure of £1029000 as the proceeds of the sale. Ignoring the rent, he has only made £1500 (£1029000 − £1027500) as a result of the transaction, a return of a mere 0.15% on money invested.

Although shares in property companies may rise or fall very rapidly, the advantage of holding shares is their liquidity; they can normally be easily sold at relatively short notice. The very volatility of the share price offers rapid trading conditions which simply do not exist in the direct property market although the development of the unitized property market will give investors the opportunity of trading their interest in a property.

2. One of the critical problems of property ownership as an investment is knowing which property to purchase. Such factors as location and building design can significantly affect the viability of the investment. This problem is particularly acute for an investor first entering the field of property investment. If an investor owns a portfolio of say four buildings, he may find that one of his buildings does not increase in value initially due to the structure of the lease or being poorly located. This one building may significantly affect the performance of the entire portfolio, which, as a result will not keep up with the general trend of the property market. Owning property shares is a way of spreading the risk, particularly if the property company concerned owns a wide range of property assets. Another way is to purchase property units.

3. Technological change may make buildings become out of date and hence lose their full investment potential. For instance, almost all new office buildings are now designed with a carpeted floor and a suspended ceiling. Buildings constructed during the 1960s without those refinements no longer command the best rents. Industrial buildings also suffer from technological obsolescence. Owning shares in a property company which owns modern buildings relieves the investor of the problem of when to dispose of a depreciating building.

4. Changes in transport investment and management can have significant effects on property. For instance, a decision to cease routeing fast trains through a town, or conversely stopping fast trains at a station previously used by slow trains, may dramatically affect property values, particularly office rents. Likewise, the opening of a new motorway or bypass can also affect property values, especially the value of industrial buildings. Whilst it is in the art of good property investment to be aware

of and respond to these changes, they sometimes have unforeseen adverse effects on the investment performance of property.

5. Town planning decisions in the UK can also influence property values. Perhaps the most obvious example is the result of the pedestrianization of a shopping street. Retail property within the pedestrianized zone often increases in value. The converse is true of shops outside a new pedestrian area; the capital value of such premises may fall back dramatically over a period of a few months and wipe out the increase in capital value which has increased steadily over 10 years or more. The opening of a new shopping centre may have a similar effect on existing property values. Owning property shares, which are usually readily marketable, does not require the investor to be involved in factors such as town planning decisions which may affect property values.

To avoid the problems which may arise with direct property investment it is increasingly important to undertake regular portfolio analysis and indulge in active portfolio management.

3.9 Investment portfolio analysis

No review should be made of the rationale behind direct property investment without considering investment portfolio analysis. This is the measurement and historical review of individual properties and groups of properties within an investment portfolio. It has an increasingly important role to play in the process of active property investment portfolio management. A prerequisite of and minimum yardstick for property performance measurement is a regular valuation of all properties.

From valuation data and from details of a portfolio's cashflow (both rental and capital), it is possible to measure:

1. Capital returns;
2. Income (rental) returns;
3. Total rates of return;
4. Rental growth.

These figures can be calculated over the short term (one year) or over a long period (since date of purchase) and the results can then be presented for an individual property, groups of properties or for the portfolio as a whole.

The most valuable part of portfolio analysis is the review of the portfolio in relation to the property market as a review of each property aids management decisions in relation to:

1. Future investment strategy;
2. Restructuring a property's tenure to maximize future performance;
3. Disposing of property which has passed its performance peak;

4.   Regeneration performance by refurbishment expenditure.

Portfolio analysis should be undertaken independently by analysts who can draw upon their extensive day-to-day knowledge of the property market. This regular, independent and systematic appraisal of a portfolio can often uncover overlooked opportunities.

The analysis will benefit from comparing the portfolio with a regular monitoring of rental levels, investment yields and the level of supply and demand in different sectors of the property market. However, a portfolio review requires a close liaison with a fund manager so that a balanced view between an independent analysis and a detailed involvement is achieved. With the aid of sophisticated computer software it is possible to analyse an investment portfolio of fewer than ten properties to one containing several hundreds.

3.10 Conclusion

The essential ingredient to direct property investment is understanding the market and knowing how to measure investment performance over both the short and long term.

References

Darlow, C. (ed.) (1983) *Valuation and Investment Appraisal*, Estates Gazette, London.

Darlow, C. and Morley, S. (1982) 125 years of property investment, *Estates Gazette*, 30 April.

D.T. & C. (annual) *Money into Property*, Debenham, Tewson and Chinnocks, London.

Frazer, W.D. (1984) *Principles of Property Investment and Pricing*, Macmillan, London.

I.P.D. (annual) *The Annual Review*, Investment Property Databank, London.

McIntosh, A.P.J. and Sykes, S.G. (1985) *A Guide to Institutional Property Investment*, Macmillan, London.

Plender, J. (1982) *That's The Way The Money Goes*, Andre Deutsch, London.

4
Finance for property development

There is a wide range of finance available for industrial and business space property development. Before considering these sources, it is necessary to distinguish between short-term and long-term finance.

Traditionally there has been a difference between short-term funding, where risks are high and returns are also high, and long-term funding where the risks are meant to be less yet the investment yields are less due to investors expecting rental growth in the future which compensates for the reverse yield gap. This reverse yield gap is the difference between the yield on long dated government stock and the lower yield of a particular investment. Property companies with long-term borrowing and equity finance are likely to prosper at the expense of those companies which fund each property development deal by deal and have sometimes to borrow expensive short-term money.

With the decrease of both inflation and rental growth coupled with the general increase in the level of investment yields during the early 1980s, the distinction between yields for long-term funding and short-term interest rates became less than during the 1970s. New methods of funding property development evolved in this period. At the same time there was a change in the real interest rate applicable to the financial markets. During the 1970s, when there was high inflation, real interest rates were often negative. Since 1982 there have been more than five years of positive real interest rates. Long-term property investment yields have generally increased, many to the same level as short-term interest rates particularly in areas of the industrial property investment market. During 1988 there were signs of investment yields falling once again as a result of very strong rental growth recorded in the office and industrial markets.

4.1 Taxation and taxation allowances

Taxation is only a source of money in the sense that the tax structure nowadays is less punitive than it used to be. Corporation Tax has fallen in recent years from 52% to 35%. This has assisted a number of property companies who are now able to be more active in the market without incurring adverse tax penalties. Lower inflation and the indexation of Capital Gains Tax, and the abolition of all CGT before 1982, has also assisted this course although the imposition of Value Added Tax on building works, has marginally increased the tax burden for some companies. The recent removal of all gains tax before 1982 has lifted the tax burden and has been especially beneficial for some of the more established property companies.

Taxation allowances particularly are also a substantial source of quasi-finance. Since 1985, the very substantial Industrial Buildings Allowance and Plant and Machinery Allowances which did exist have been removed. However, Industrial Building allowances still exist at 4% on a straight line basis over 25 years while Plant and Machinery Allowances still exist at 25% on a reducing balance basis. In Enterprise Zones 100% of capital expenditure is an allowance against corporation and/or Income Tax until 1992. For this reason special funding arrangements are applicable in certain parts of the country. Both personal and corporate monies have been invested in the London docklands particularly in recent years and recently there has been an advertizing campaign selling property investments in other Enterprise Zones throughout Great Britain. For the high tax payer, such investments can be very beneficial although the reduction of the higher rate of tax to 4% in 1988 has made such investments less beneficial than in the past.

4.2 Assisted areas

In addition to taxation allowances, a number of parts of Great Britain still offer government grants and incentives. Although these have been reduced in recent years, they are still substantial and such organizations as English Estates have made good use of this source of finance. The system of Urban Development Corporations, which have been increased in number in recent years, is an example of this type of public/private sector partnership.

There is also City Grant. This follows from the Inner Urban Assisted Areas Act of 1978 which was revised in 1982. It is essentially a pump priming source of public finance where the central government originally provides additional finance to enable certain projects to be funded where they would not otherwise have been possible. City Grant is an amalgamation of Urban Development Grants, Urban Regeneration Grants and Derelict Land Grant although these earlier grants were administered through local authorities. In addition there are various loan guarantee schemes where investment loans are guaranteed up to 70% of the original money funded. All these aspects are covered in greater detail in Chapter 9.

4.3 Private sector finance

The following sets out the various sources of funds in the private sector which are available to those contemplating property development. First it is necessary to understand the difference between recourse and non-recourse finance.

Over the last ten years, the loans outstanding by the banking sector to the property world have increased dramatically. In a matter of five years to 1988 loans increased from £3 billion to more than £18 billion.

These loans are mainly of two types, recourse and non-recourse. A recourse loan means that the bank or lending institution has complete recourse to the finances of the organization to whom it is making the loan. By taking a charge over the company, if there is default in payments or if the company goes into liquidation, the lender has a claim on the company's assets before other claims are settled. Interest due on the loan for instance has a prior claim over dividend payments to shareholders and other forms of equity.

A non-recourse or 'off balance sheet' loan restricts the claim of the lender to a particular project or asset. The majority of limited non-recourse loans are arranged using a single purpose company. This allows the lender to take all the necessary legal obligations and charges over the single purpose company, without restricting the business of the parent company. It has become widely accepted that for limited or non-recourse loans this method is the most efficient both in practical and legal terms.

A limited recourse loan is, as the name suggests, where the lender has recourse to the project concerned and, to a limited extent only, to the finances of the development company normally through various guarantees. In reality there are very few completely non-recourse loans made as there is almost always an element of limited recourse payment involved.

To make a loan, a bank will undertake considerable analysis before making a decision. The first is to consider the accounts of the company wishing to take out a loan. The problem is that a balance sheet relates only to one moment in time and the figures it contains are open to interpretation. The notes to the accounts may he helpful but by looking at several years a picture of the finances of the company can normally be put together. Looking at the accounts, the name of the auditor, the date of the balance sheet and the audit date may be helpful. The two dates should not be far apart but there may be important credit factors influencing such a delay.

The current assets and liabilities are important, particularly the liquidity of the assets if they need to be sold quickly. However the asset may already support, as collateral, hire purchase agreements, other secured loans, debentures and mortgages.

The three most important tests are therefore: the liquidity of assets, the profits of the company and the capital stake the company is prepared to put into the project. This latter element is often described as the equity element as it contains more risk. Returns on this tranche of the investment are only achieved if all the debt payments have been met successfully.

4.4 Ratio analysis

Ratio analysis, although a straight forward mathematical assessment, helps build up a picture of the company and help a bank decide on the nature of a loan it can grant for development. The various ratios to be considered are:

1. Return on capital is found by dividing the pre-tax profits by the net capital resources. Pre-tax profits is the net earnings after all charges but before taxation, dividends and appropriations. Net capital resources is the share capital, reserves, and deferred taxation which does not require immediate payment. The ratio measures the efficiency of the company and is used to gauge volatility over time, how it compares with other companies and whether it is keeping pace with inflation. The problem is it relates to balance sheet values, which may or may not be realistic; the market value of the same assets may be very different.

2. Liquidity ratio is the current assets divided by the current liabilities. This measures the Company's ability to absorb losses without causing problems, its ability to build up liquidity to meet maturing debt commitments and its ability to enter into a new capital expenditure programme. One of the difficulties, particularly with property companies and building contractors, is to assess the stock and present work in progress. Liquidity can often change very quickly, perhaps week by week or even day by day.

3. Gearing is calculated by dividing the balance sheet borrowing by the net capital resources. This ratio is an indication of capital sufficiency and whether the borrower is already relying on borrowed funds but it also indicates whether further funds could be raised. If a company is highly geared, when times are good, dividends increase significantly to the benefit of shareholders in terms of the dividend income and the capital growth of shares. When times are hard, highly geared companies find that they get into cashflow difficulties. Their liquidity falls such that they are unable to realize the value of their assets yet pre-tax profits fall also so that they are unable to meet immediate cashflow commitments. Such a scenario created the property crash in the 1970s.

Apart from the three important ratios above, of return of capital, liquidity and gearing, there are other ratios which may also help in understanding a company.

4. Leverages are found by dividing the total balance sheet liabilities by the net capital resources. This is an extension of the concept behind the gearing ratio and can show up a higher level of debt capital to equity capital than is desirable.

5. Income gearing is the profits before interest and tax divided by interest. This shows the vital relationship between profits and a change in interest rates.

6. The acid test is the current assets, less stock and work in progress as a proportion of the current liabilities. This is sometimes referred to as quick assets over current liabilities and emphasizes the need always to have a positive cashflow.

7. The capital ratio is the net capital resources divided by the fixed assets. This will indicate how much a company is contributing to its capital requirements.

8. The gross margin shows the profit before central overheads as a proportion of sales. Comparing a number of similar companies it is sometimes possible to identify those which are efficiently run and those which are not. If the gross margin is high yet the return on net capital resources is low, there may be cause for concern.

9. Finally the net margin is the pre-tax profit (excluding central overheads) as a proportion of sales. Comparing one company with another, it is also possible to judge the efficiency with which a company is managed. These last two ratios are less relevant in the property industry but may be particularly relevant for instance in the retailing world.

The simplest type of funding is where the lending only relates to the company undertaking property developments. Some companies undertake a number of investment projects at any one time.

When a new company is formed for just *one* project and there is no track record of previous financial behaviour, analysis is far more difficult. In these cases the project, the team and guarantees are important to the project. However, speculative development, even if the management have a good track record, should be regarded with caution. The wrong type of property in the wrong location and an incorrect funding arrangement can create later difficulties for everyone including the bank providing the finance. One way round this type of problem is to create a joint venture company where one party offers the financial security and the other in effect manages the project development.

4.5 Presentation of case

The presentation of any project to a bank for funding consideration is paramount. All the above ratios need to be taken into consideration before the presentation is made.

The first item of the presentation should relate to the location of the site for the proposed development. This will be considered in terms of the type of planning consent which has been granted; detailed, outline, or none at all.

If a loan to enable building works to be paid for is agreed, the banker may also permit the loan to run for an interim period of five years or more after the project is complete. Depending on the nature of the development and development company, the bank will rarely be happy for the loan to be extended over a longer term.

Having considered the market in terms of location, communications (infrastructure), supply and demand for similar buildings, the bank will also want to know whether the development is pre-let, whether there are alternative uses should the market change and whether end finance in the form of a long-term investor or owner occupier has been identified.

When considering the appraisal the bank will examine costs in relation to the building contract, professional fees, legal fees, planning fees and the margin of sale price or completed property value over cost. Building timing in relation to the cycle of the market must be considered relating to the construction programme, the cashflow forecast and sensitivity of analysis. The financial appraisal should be examined by varying the interest rates, rents, building costs, building delays, rent voids and investment yields.

Inevitably there is an element of subjective judgement but property development in some locations is more predictable than in others. The trend of the national and international economy also needs to be assessed. The critical assessment is whether, if necessary, the banker could take over the development project, complete the development and then raise a mortgage on the completed asset. For this reason the banker will usually seek a capital commitment from the developer and will normally end up to 75% of the value of the asset depending on the nature of the scheme.

Before actual payments are made, the banker may require the project manager and/or the architect to issue certificates. Even these will not always be reliable evidence that the development programme is progressing satisfactorily; in an extreme case the banker may have to rely on the courts to settle a claim for negligence!

4.6 Interest payments

Short-term loans will be charged interest calculated in relation to LIBOR (London InterBank Offered Rate) with rollovers at 3, 6, 9 or 12 month periods. LIBOR is quoted for 1, 3 and 6 months and the rate used will depend on the type of loan. LIBOR is based on a group of British and overseas banks and is adjusted daily. Fixed rates may be applicable to interim or longer loans of 5 to 10 years although interest rates may vary within a 'collar' between the 'floor' (lower rate) and the 'cap' (higher rate) which may be preset at the start. A regular payment or 'bullet' lump sum paid on completion of the development is sometimes used.

For longer-term finance, the rate may increase if the rent or income received is above a preset target rent agreed at the outset of the development. One popular concept is that of 'mezzanine' finance, a form of unsecured debt. Whilst debt finance has a first charge on a property or development, once this claim has been met, those providing 'mezzanine' finance are entitled to a higher rate of interest. Although this is a corporate line of finance it is not totally different from the concept of debenture stock or preference shares which have priority over equity interests.

4.7 Fees

Most loans have an element of fees, for instance, for initially arranging the loan or whenever there is a 'draw down' to pay an architect's certificate during the course of construction. In addition to these charges there may be trigger clauses of 'equity kickers' payable in a similar way to additional interest payment previously described, although these may depend on changing capital values rather than rental income.

4.8 Leasing finance

This type of finance is particularly applicable in Enterprise Zones where large tax allowances exist. In such situations it is necessary for the financier, often a merchant bank, to take a qualifying interest in the property to be able to obtain the taxation allowances previously referred to and then grant a leaseback to the occupier and/or funder at favourable rates reflecting the tax allowances gained.

The recently opened Metro Centre near Gasteshead is an example of such an arrangement where the County Bank were involved alongside the developer (John Hall) and the investor (the Church Commissioners). As many buildings nowadays contain valuable plant and machinery which may qualify for tax allowance purposes, the potential for such an arrangement can be considerable.

4.9 Property companies

For those wishing to go into partnership with a property company they may either enter into a tenure arrangement or form a joint venture company. Property companies as a source of finance fall into three areas; property investment, property trading and property development. Some companies also have interests in allied fields such as construction, civil engineering and, because of the property element, high street retailing. The property investment element of their activity is largely akin to the property invest-ment behaviour of the investing financial intermediaries. However, unlike the institutions, their behaviour is more speculative and may often be more concerned with short-term gains rather than long-term investment. To comply with statutory regulations, property companies must produce audited accounts.

The principal disadvantage of investing in property companies is that they suffer both Capital Gains Tax and Corporation Tax. However, the precise tax situation of individual companies will depend upon the nature of the business as defined in their respective Articles of Association. Advance Corporation Tax is incurred during the year whilst the balance is payable at the financial year end. The dividend payable to shareholders is also taxed, although the company may offset ACT against the income tax which it is obliged to deduct automatically from dividends prior to their distribution. Tax exempt shareholders, therefore, have to reclaim their tax which has been paid at the basic rate of income tax. As with other property investors, property companies may also be liable to stamp duty and VAT on their property investments.

Property companies are not really directly comparable with other forms of property investment, as the share price and performance is determined not only by the underlying assets, but also equity market sentiment. Property shares can be very volatile partly as a result of their speculative nature, the speed with which they can be traded compared to direct property and the fact that a company may be financially geared to a greater or lesser extent. Balance sheet gearing (i.e. the total borrowings compared to shareholders' funds) varies greatly between companies, and generally the higher the gearing, the more speculative and volatile the share price.

Whilst gearing has the effect of increasing the shareholders' funds at a greater rate than increases which may occur in the market value of the assets held by a company, there are two obvious risks. Firstly, if property values fall, the shareholders' funds will fall at a greater rate. Secondly, where debt is at variable rates, if interest rates increase, this may rapidly reduce, or even eliminate any profit shown by the company.

The commonly accepted measure of the worth of a property share, and to a lesser extent especially investment companies and trading companies, is often its net asset value (NAV), calculated by dividing the shareholders' funds (i.e. net worth of the company after paying off all debts) by the number of shares in issue. This may be slightly more complicated in the cases of companies with convertible loan stocks or other share options.

The calculation of the NAV may well entail the estimation of the market value of the properties on the basis of a current valuation, as the value in the balance sheet may relate to earlier years. The discount which the share price exhibits compared to the NAV is the most commonly used method of estimating the relative cheapness or expense of a particular stock. Dividend income is usually a secondary consideration. For companies with a large development programme, estimation of the NAV may prove problematical in view of the vagaries of valuing developments in progress and their speculative nature.

4.10 Investment trusts

Investment trusts are a long established way of enabling an investor to spread the risk of his investment. They are joint stock companies which invest in equities, debentures and preference shares of other quoted securities. They offer the investor an equity share in a fund which owns a variety of securities both in the UK and overseas.

The investor in an investment trust company normally seeks both income and capital appreciation. An investment trust cannot hold more than 15% of the value of its investment in unlisted company securities and investment in any one company must not exceed 10% of the value of the trust. Investment trusts do not often own property investments and must hold 15% of its assets as cash.

There are special CGT arrangements which reduce the liability to this tax. However income is subject to Corporation Tax. Dividends are taxed like other shares requiring exempt investors to reclaim the tax from the Inland Revenue. As with financial intermediaries, investment trusts are liable to stamp duty and VAT in certain situations.

4.11 Rights share issues

Over the last year or so property companies such as Rosehaugh/Stanhope Developments involved with the Broadgate Development in the City of London have raised millions of pounds in rights issues. During the last few years several other names involved with the property and building industry have also raised finance through this method including McCarthy and Stone plc and C.H. Beazer. Such a method of raising finance is only applicable for the better run property companies and may be particularly relevant for those involved with trading rather than investment strategies. Essentially, existing shareholders are offered the right to extend their shareholding by being given a preferential opportunity to purchase additional shares. Raising finance through this route will, however, depend on the state of property shares and the equity market generally.

4.12 Debentures, mortgages and bonds

There are an increasingly wide variety of debentures now being issued on the market. Over the last few years such companies as Land Securities have raised £100 million using a 9.5% yielding 2007 bond. Land Securities have also raised many millions using 10% first mortgage debentures. In 1988 Slough Estates plc raised £150 million incorporating a 'rolling put' option for investors and a 'rolling call' option giving rights of redemption to the property company.

There is also the concept of *Deep Discount Stock* where the investor can earn, not only interest on the stock but receives an additional capital payment at the maturity date to reflect the shortfall on interest payments from a full open market rate. In recent years such companies as Tesco plc and Beazer Homes have raised money through this source. During 1985, for instance, Safeway also raised funds using this method issuing £100 million, where the value of the stock was to rise in line with the rental value of a notional portfolio of supermarkets. Wickes have also in the past set up an 'Off balance sheet' property holding company using this technique. Burtons used this method for raising £70 million based on the value of a portfolio of Debenham Stores.

This method of raising finance has become an integral part of the recent move to securitize property. It enables companies to raise finance based on the value of properties without actually selling their assets which was a common practice in the 1960s and 1970s.

Mortgage loans are another type of finance available. The concept of the 'equity kickers' is not dissimilar to discounted stock in that the mortgagee receives a lower initial return on the loan but is paid a capital sum at the end of the loan period, an equity kick, which may depend upon the capital gain of the asset upon which the mortgage or loan has been made.

In addition to the above types of debenture issues one should not forget the other schemes around which are generally corporate lines of finance not placed on the open market. In the last few years a number of property companies have raised money by issuing Sterling Commercial Paper which is a direct contract between the borrower and lender without a banker inter-mediary.

Sterling Commercial Paper is not dissimilar to Euro-Bond Issues which have been common for a number of years and are often issued in dollars. A number of property companies have also raised Euro-Dollar loans during the last few years including Land Securities, MEPC, Slough Estates and Wates. There has also been money raised in Deutschmarks by Hammersons.

4.13 Joint ownership

This type of financial arrangement has been taking a new course in recent years. There are a number of reasons for this development.

First, there are very few institutional investors in the investment market prepared to fund development schemes worth more than say £50 million. When one considers that many of the proposed shopping centres and office buildings in the City of London have an asset value in excess of £100 million, one can see immediately that some form of securitization or unitiz-ation may be appropriate.

Secondly, there are a number of developers who wish to sell on completed schemes. One way to do this is to unitize their developments and sell on the interest in tranches.

Thirdly, there is a need for greater liquidity in the property investment market. It often takes many months to invest or disinvest in the direct property investment market. The concept behind unitization may enable people to purchase or sell units in a particular building on an hour by hour basis.

Fourthly, by creating a unitized market, small investors can spread their investment risk across a variety of property investments by acquiring only a small holding in each building.

4.14 Unitization

Some forms of unitization have been with us for some time. First, there is the traditional 'trust for sale' arrangement such as the sale by part of the Stoneborough Shopping Centre in Maidstone by Prudential to the Shell Pension Fund a year or so ago. One party has management rights yet both parties have to agree on any material change with the way the asset is managed such as the recent announcement that the centre is to be refurbished.

Secondly, there is the concept of securitization which is the creation of different layers of interest within a single building, shares for which may be acquired in a stock market. In the USA this concept was applied to the New York Rockefeller Center and has been widely used since. In the UK a building which has been securitized is the office building at Billingsgate in London. This securitization often referred to as SAPCOS (Single Asset Property Companies) involved three layers:

1. *A deep discount bond* which is basically a first charge on the property and its rental income. For Billingsgate £35 million was raised from this first layer by S.W. Berrisford, the developers, who are required to repay £52.5 million in May 2006. This is a deep discount mortgage debenture bond because the repayment of capital at face value at that time will be larger than the original sum loaned creating an eventual capital gain.
2. *The preference of ordinary shares* raised £25.8 million at Billingsgate and provided the purchaser with a yield of 5.9% gross of tax. Because of Stock Exchange rules at the time, these shares were quoted on the Luxembourg Stock Exchange. The owners of the shares were entitled to 30% of all increases in income and 30% of the capital appreciation at the redemption date.
3. *The ordinary shares* remained in the hands of S.W. Berrisford, the developer. No income of dividend will be payable until the rent review when they hope to receive 70% of the uplift in rents above the rent now passing. All running expenses of Billingsgate City Securities plc are paid for by the ordinary shareholders.

Since Billingsgate was securitized, London and Edinburgh Trust set up a new wholly owned subsidiary company initially worth £53 million called LETINVEST based on this financial formula. The new company concentrates its efforts on investing in a number of high yielding property assets.

Apart from 'trust for sale' and 'securitization' there is also the concept of 'unitization'. There are currently two vehicles, first is PINCs, otherwise known as Property Income Certificates, which work within existing legislation and involve the passing of rent through a management company and financial intermediary to the investors. Control is operated through ordinary shares in the management company. The contract therefore relates to the payment of rent, not an interest in the land. Tax is not deducted at source but all income is received gross by the investor who then makes his own tax arrangements.

The second unitization vehicle is the 'single property ownership trust' (SPOT) which has been developed by John Barkshire of Mercantile House. The Financial Services Act 1987 theoretically created a new legal framework within which single property ownership trusts may operate. Each unit is a part legal title to the property and the trust deeds provide details of the management of the investment. Rents are passed gross to the unit holders to make their own tax arrangements. The Finance Act 1967 should make the units tax transparent by giving the Treasury the power to exclude the units from the statutory code for unauthorized unit trusts. However, this has been the main obstacle in the way of launching the SPOT market.

4.15 Financial institutions

During the 1950s, 1960s and 1970s when there were restrictions on bank lending, many property companies turned to life insurance companies and later pension funds as a source of finance. The institutions are part of a group of organizations known as financial intermediaries. In basic terms, these intermediaries perform the function of transferring funds from persons or organizations with a monetary surplus to those requiring additional monies in order to undertake investment or expenditure.

An example of an intermediary is a pension fund. Throughout an individual's working life contributions may be made, on a monthly basis, towards a pension fund which invests the savings for withdrawal at a future date following retirement. The size of the lifetime contribution and, at a later date, the size of the pension, will depend on efficiency of the pension fund and the effectiveness of the investment strategy used by that pension fund.

High Street Banks, Merchant Banks and Building Societies are also financial intermediaries. Although the financial investing institutions such as life insurance companies and pension funds are financial intermediaries, there is an important difference. They are able not only simply to provide loans and mortgages to borrowers, but are also able to invest their funds for long periods of time, often many years. The staggering growth of these institutions has gradually replaced individuals as the main investors in the UK. The social, economic and political effects of these so-called leviathans of the financial world have only recently been appreciated.

One such area of change is the way many commercial and industrial buildings in the UK's cities are now financed. Many buildings now have to be located, designed and constructed in such a way as to be acceptable to these financial organizations. The hidden hands of investing institutions are, to some extent, dictating the shape of the UK's urban areas. Since 1982 the sums of money allocated to property investment has declined at the same time as loan to property companies by the banking sector have increased. However, in recent years, in excess of one billion pounds has still been placed in the property market by the investment institutions. There is now over £36 billion invested in the property market by institutions and there is no doubt that they will continue to be a major force to be reckoned with in the financing of future property development.

4.16 Insurance companies

There are over 800 insurance companies in the UK, although only about 200 are registered with the British Insurance Association. Their aim is to spread risk between organizations and individuals over periods of time although they are required to demonstrate certain solvency margins to the Department of Trade and Industry.

Insurance falls into two principal categories; life assurance and general insurance. This second category includes fire, marine, motor and building insurance. General insurance is normally carried out on a year-by-year basis with premiums being reviewed annually. When large insurance claims are made, assets must be quickly realized to enable claims to be met. Only a limited amount of general insurance finance is available for long-term investment purposes.

Life insurance can be divided into three principal categories: term assurance, whole life cover and endowment assurance, insurance on a life is taken out on a year-by-year basis in a similar way to a general insurance policy.

Whole life with an endowment involves paying a regular premium for a period of years. At the end of that time, assuming the beneficiary is still alive, a payment is made on the maturity date. If the policy is with profits, the assured will receive a bonus sum. The profits or bonus will depend on investment acumen of the insurance company. Although some of the companies are publicly quoted on the stock market, many others are mutual companies. Within this latter group, the policyholders are, in effect, the shareholders of the company and mutually benefit from the success of the company's investment strategy, although the opportunities to question or understand the strategy are very limited.

Annuities provide policy holders with a regular income normally following the payment of a premium or a number of premiums over a period of years. Once again, the regular income received by the policyholder will depend on the company's skill, both in efficiently running the company and also in shrewdly investing the policyholders' premium or premiums.

Over the last fifteen years, within insurance companies, two new types of vehicles for savers have emerged; individual and managed pension funds and unit-linked life assurance.

The individual and managed pension funds originally grew as both the self-employed and employees became more aware of their financial future. It became axiomatic that a secure pension was part of a rising standard of living. As a result, individuals wished to take out personal pension plans and the insurance industry responded to this demand.

The managed pension schemes were developed to enable companies who were managing their own pension plans to invest in a sound investment portfolio. A variety of company pension fund trustees now purchase units in these funds. The funds in many ways are very similar to the unit trusts previously briefly described. The principal difference is that the managed funds are often able to use the long-standing investment experience of the parent insurance company and the funds are gross funds. Income and capital gains are not taxable as they have pension fund status. The growth of the unit-linked life assurance may be partly attributed to savers being able to gauge the performance of their savings. The premiums are entitled to tax relief, yet the money which accumulates is unitized in a similar way to a unit trust. The saver can purchase units in a general fund or can specialize by investing in a particular fund such as the equity fund or the property fund. After the saver has paid premiums for a statutory minimum number of years to qualify for tax relief on the premiums, he can redeem some or all of his units. A life assurance policy which is unit linked is not dissimilar to an endowment with-profits life policy. The main difference is in being able to observe the success or otherwise of the savings and flexibility of offers to the saver.

4.17 Pension funds

Pension funds are very similar to the annuity business of life assurance companies; contributions are made to the fund by members over a period of time, which then become pension payments to the employee following his or her retirement.

An individual participating in a private pension plan will receive a pension dependent on the successful organization and investment strategy of the fund. These funds are often managed by insurance companies. These schemes and all other pension schemes have to meet certain statutory minimum standards of return to pension fund members as set out in the Social Securities Pensions Act 1975.

The majority of pensions are paid by employers to employees on the employees' retirement. The payments are normally paid monthly until the former employee dies. The employee is therefore, in effect, a saver through-out his working life, which may be in excess of forty years. After retirement most employers will guarantee an employee (now a withdrawer) a certain pension. In private industry this will be a percentage of the employee's retirement salary as provision is made for the pension to be dynamic. In other words, each year the pension will grow to compensate for inflation, but will rarely keep pace with inflation.

In the public sector, such as the Civil Services, the pension scheme also provides a pension based on a percentage of the employee's retirement salary. The pension is then reviewed each year and, in the majority of cases, increased in line with inflation.

Both private industry and the public sector guarantee employees a certain pension. The pension fund is often administered by trustees. It is therefore up to the trustees to invest the pension contributions as shrewdly as possible, to enable the fund to meet the commitment of future pensioners. Any shortfall between pension payments and investment returns on contributions invested has to be made up by the trustees. In the case of private industry, the company itself has to make up the shortfall although in recent years the investment returns have often resulted in the reverse being appropriate. Investment income has outpaced the sums needed to meet pension commitments.

In the case of the public sector, the taxpayer indirectly makes up the shortfall to bring the pensions in line with inflation. In both the public and private sector, the investment strategy of the trustees is vital. Such invest-ment strategy nowadays may include purchasing investment property.

4.18 Owner occupiers (internal funding)

We should not forget that, in recent years there has also been a drift back to the owner occupation of office and industrial buildings rather than renting from an investor. There are both advantages and disadvantages to this course of action but it is becoming popular with the small office market where companies can either use their own building as collatoral for raising further finance or they take the building into their own pension fund and then lease it back to themselves. The rent paid can be offset against taxable income by the occupier whilst the pension fund receives the rent gross of tax.

A similar trend has been noticeable with larger industrial and office buildings. Such changes as the indexation of Capital Gains Taxation has meant that, for a company to own the buildings it occupies is more advan-tageous than before 1983. Owner occupiers are also able to benefit from Plant and Machinery Allowances which they can offset on a reducing balance basis of 25% against Corporation Tax, while Industrial Building Allowances of 4% on a straight-line basis may also be applicable in some cases.

As Corporation Tax has fallen there has been less incentive to offset rents paid against taxable income but it is the removal of controls on credit within the economy which has changed attitudes most. Companies are no longer discouraged from locking up capital in property which was often the case during the credit squeezes of the 1960s and 1970s.

The trend towards owner occupation is linked in many cases with the use of lease finance previously discussed. The occupier buys a term in years for a capital sum raised through a finance company and then makes repayments to that finance company who is able to claim tax allowances.

Many overseas companies now purchasing freehold interests in the UK can offset the depreciation of the investment against tax over a number of years, something which is not permitted by present UK taxation law.

4.19 Conclusions

All those involved with property development must be clear in what they are developing. The development must be in the right place and designed suitably for the tenants of that particular market. Tenant demand must be measurable and, if a developer can arrange a pre-let, he is clearly going to find it easier to finance the scheme.

Occupiers, developers, investors and local authorities should explore and be aware of all the methods of funding property development now available. Difficult areas of the property market will suit different methods of financing. Increasingly investors have to be aware of the problems of obsolescence and refurbishment expenditure. These have to be brought into the financial equation so that it is possible to regenerate the property's investment value by periodic expenditure.

In recent years a few firms of chartered surveyors have set up financial services companies. These fall within the Financial Services Act 1986 and are able to offer a much wider range of financial advice than was previously available. The firms are also able to call upon their commercial agency knowledge, valuation skills and management expertise to offer the 'one stop shop' for all types of property finance.

References

Cadman, D. and Austin-Crowe, L. (1984) *Property Development*, 2nd edn, E. & F.N. Spon, London.

Catalano, A. and Huntley, J. (1987) Bank financing. *Estates Gazette*, 12 December.

Chua, K. (1987) Sources of funding. *Architects Journal*, 14 January.

Clark, W. (1987) *A Banker's View of Funding*, Proc. of Conference on Getting Development Funding Money, Midlands Study Centre, April.

Darlow, C. (1987) Development finance – Finding the funding. *Estates Gazette*, 14 February.

Darlow, C. (ed.) (1988) Valuation and Development Appraisal, 2nd edn, *Estates Gazette*, London.

Evans, P. (1987) Property finance – banking on property. *Estates Gazette*, 2 March.

Evans, P. and Jarrett, D. (1988) Property investment non-institutional funding. *Estates Gazette*, 11 June.

Gibbs, R. (1987) Raising finance for new development. *Journal of Valuation*, **5**, 343–53.

Healey & Baker (1982–88) Quarterly Investment Reports, Healey & Baker.

Orchard-Lisle, P. (1987) Financing property development. *Journal of Valuation*, **5**, 354–68.

Parry Wingfield, M. (1987) Tax relief and property finance. *Journal of Valuation*, **5**, 390–400.

Part Two
PROPERTY COMPONENTS

5
Traditional light industrial development

The concept of 'light industry' arose after the Second World War when it was realized that new industries were growing which were 'footloose'; they did not need to be tied to sources of raw material, energy or such factors as ports and seaborne transport. The 'footloose' concept has often been misunderstood. The facts are that all modern industry is very location conscious and in reality the concept of 'footloose' does not exist. What has changed are the criteria used by industry to decide upon one location rather than another.

5.1 Classification of planning uses

Before examining location criteria it is necessary to understand the concept of a light industrial building. Until recently the Town and Country Planning (Use Classes) Order 1972 set out a number of the use classes. Light industrial fell within Class III and was defined as an industrial activity which might be appropriate within a residential area.

In addition to Class III, Class X related to a building for the 'Use as a wholesale warehouse or repository for any purposes'. Many investors and developers still group these two types of building use together and refer to them as 'industrial property'.

The Town and Country Planning (Use Classes) Order 1987 has slightly changed the rules of the game. The new Class B1 has absorbed the old Class III but also permits a wider use. Such a building may also be used for office purposes by an industrial manufacturing company and increasingly by the financial services and/or professional office users.

The new Class B8 replaces the old warehouse Class X but now includes the use of land for storage purposes as well as buildings. Over the last ten years a number of areas previously allocated to light industrial development

Table 5.1 Guide to business space rental values (as at December 1988)

Location	Traditional	Detached standard HQ	Mixed use business premises	Out of town office
	10% offices	25% offices	50% offices	100%
	£ psf	£ psf	£ psf	£ psf
Bedford	3.50	3.75	5.50	7.00
Birmingham/Coventry	3.75	4.25	8.00	12.00
Bristol Area	4.25	5.25	8.50	12.00
Cambridge	4.50	5.50	10.50	15.00
Cardiff/Newport	3.50	4.00	5.00	8.50
Colchester	4.00	4.65	6.25	8.00
East Berkshire	6.00	7.50	12.00	20.00
Edinburgh	2.50	3.00	3.50	
Glasgow	2.25	3.00	3.50	
Leeds	3.50	3.65	5.50	8.50
London Enterprise Zone	7.25	10.50	15.00	20.00
M25 South-West	7.50	8.25	15.00	18.00
M25 West	7.25	9.00	14.00	20.00
M25 North-West	6.25	7.50	12.00	15.00
M25 North-East	5.50	6.25	10.50	12.00
M25 South-East	5.25	6.50	11.50	15.00
Manchester/Warrington	3.25	3.75	6.50	9.50
Milton Keynes	4.25	5.10	9.50	12.00
Northampton	3.75	4.25	6.50	8.00
Norwich	3.50	3.75	5.25	8.00
Oxford	5.25	6.00	9.00	9.00
South Hampshire	4.25	5.25	8.00	8.00
Swindon	4.50	4.75	7.00	8.00

Source: Healey & Baker

have been taken up by the development of retail warehouses, sometimes in the form of a retail warehouse park, and on other occasions by the construction of a large free-standing foodstore with a gross square footage of more than 50000 sq. ft. During 1987, as industrial rents in south-east England started to rise again the reverse started to happen in a few locations.

Within the new Class B1 there are a range of different types of building in the present market which are broadly classified as low-tech, mid-tech and high-tech. There are also the concepts of seedbed or nursery units, science parks, research parks and business parks. The nomenclature can be confusing and is often no more than a technique to advertise a new estate.

The high-tech concept is often confused in the market. Architects often refer to a high-tech building as one which is **constructed** with the most advanced materials and using the most advanced techniques. The property market, on the other hand, when it refers to high-tech buildings primarily refers to the type of **user**; a company associated with advanced technology production techniques is the occupier. This might include an electronic or computer company or a pharmaceutical company. Chapter 6 considers high-tech/business space.

5.2 Low-tech and mid-tech buildings

This chapter is primarily concerned with the low-tech–mid-tech concept. These are buildings which are of traditional construction with more than 6 m floor to eaves height suitable for a range of industrial uses – perhaps even warehouse activity. Low-tech buildings normally have a low level of office content, perhaps as low as 10%. The units on some estates are part of a terrace arrangement.

Mid-tech buildings are not dissimilar to low-tech buildings. Essentially they are traditional industrial buildings which may be fronted with an office style building where the office content may represent around 30% of the total area. As with low-tech buildings the design in many cases will be flexible permitting the office content to be extended should the tenant so wish.

Mid-tech buildings sometimes are free standing (not part of a warehouse terrace) and permit a variety of uses within the same building. The 'service' type building may be occupied as a regional headquarters by companies who use them as their local office premises, undertake maintenance and repair of equipment, provide a quasi showroom facility and use part of the premises for warehouse style storage. This mid-tech concept overlaps with the high-tech/business space concept discussed in Chapter 6.

5.3 Seedbed/nursery unit

Small seedbed or nursery or workshop units have been successfully developed in a number of locations. Outside south-east England the rental value and security of rental income is often insufficient to attract private capital alone. Public sector finance, sometimes with public sector active involvement is often needed for such units to be constructed (Chapter 4). Places like Milton Keynes and Warrington led the property market at the start of the 1980s. In the private sector there have been successful schemes such as the Teddington Business Park, south-west of London, built by Wallis International Development containing 14 800 sq. ft units.

The key to the success of these units is to provide small units perhaps of 500 sq. ft for companies to start up. Seedbed/nursery units normally have a concrete forecourt and a roller shutter door and cater for lower mid-technology operations. They appeal to a different sector of the market compared to science parks discussed in Chapter 6.

Table 5.2 Sample locations for industrial land values

Region and town	1977 ind. land values	1988 standard ind. site values	1988 high-tech/B1 site values	1988 non-food retail w/house site values	1988 food superstore site values
South East					
Bedford	£50000	£325/350000	£400000	£900000	£1250000
Brighton	£175000	£500000	£1m	£1300000	£2m
Colchester	£50000	£300000	£400000	£700000	£1250000
Crawley/Gatwick	£100000	£750000	£1m	£1200000	£2m
Heathrow	£100000	£800000	£1200000	£750000	£1750000
Reading	£150000	£600000	£1m	£850000	£1750000
Southend on Sea	£35/40000	£250000	£350000	£750000	£1250000
London (S)	£175000	£500000	£1250000	£1200000	£1750000
London (W)	£100000	£850000	£1250000	£1200000	£2m
London (N)	£90000	£500000	£700000	£800000	£1500000
London (E)	£75000	£450000	£650000	£800000	£1500000
Maidstone	£500000	£350000	£500000	£850000	£1250000
South West					
Bristol	£50000	£200000	£350000	£700000	£1250000
Exeter	£50000	£150000	NONE	£650000	£1250000
Plymouth	£35000	£75000	NONE	£700000	£1250000
East Anglia					
Cambridge	£40/45000	£325000	£400/450000	£1m	£1750000
Peterborough	£25/39000	£275/300000	£300/350000	£750000	£1m

West Midlands					
Birmingham	£55000	£100000	£250000	£750000	£1500000
Stoke on Trent	£25000	£60000	NONE	£500000	£1m
East Midlands					
Derby	£30000	£90000	£100/110000	£600000	£1m
Northampton	£35/40000	£250/300000	£375/400000	£600000	£1250000
North West					
Manchester	£30000	£100000	£150000	£550000	£1m
Preston	£15000	£50000	NONE	£550000	£1m
North					
Newcastle upon Tyne	£10000	£65000	£100000	£500000	£1250000
Yorkshire and Humberside					
Leeds	£25/30000	£100/120000	£120/150000	£500000	£1m
Sheffield	£25000	£40/70000	£75/90000	£450000	£1m
Wales					
Cardiff	£35000	£100000	£250000	£700000	£1250000
Wrexham	£10000	£60000	NONE	£400000	£750000

Source: Healey & Baker.

5.4 Investors' attitudes

As a general statement investors only purchase low- and mid-tech industrial property which is **not** entitled to a government grant or benefit from enterprise zone or free port status. They are seeking property which will provide not only an initial yield (in the form of rent) on capital invested but the prospects of rental growth over a twenty year span or maybe more.

The exception to this statement is that a few investors in recent years have regarded industrial investment more like a debt financial deal. As long as the income (rent) more than meets debt interest charges the investment is attractive regardless of whether there is likely to be rental growth. Some institutional investors acquire such investments, rather than high yield gilts, because a property acquisition enables the investors to enjoy rental growth in the future or the chance to sell on to an adjacent land owner or even the possibility of generating a windfall capital gain.

The key to this investment strategy is that the initial rental income **must** be secure. If the rental (investment) yield is 10% or more on capital it is essential that the income continues to flow. For this reason a newly formed private company is unattractive. Such investors look to the lessee's covenant to be a sound public limited company or a public sector oganization, such as a local authority or development corporation. Public/private partnership financial arrangements can also be the key to redeveloping many inner city locations (Chapter 9).

For each of these types of industrial property there are different locational and market characteristics. This is despite the fact that the regional pattern of all industrial and warehouse investment property tends to be similar to the regional pattern for office investment property. There tends to be a stronger market in the southern half of the UK with particular emphasis on the Home Counties and Thames Valley. In this respect there is also an inverse correlation between the level of unemployment and the level of rent; the higher the unemployment the lower the rental value of industrial buildings tends to be.

There are, of course, exceptions to this general rule. It is possible to find areas or towns where there is an emphasis on encouraging industrial development, but the development of warehousing may be restricted as a result. This localized artificial market situation may cause rental levels of warehouse property to rise despite the vast number of empty industrial buildings.

The growth of road transport, largely at the expanse of rail transport, changes to bulk handling cargo at ports and airports, the changes in distribution methods in the retail sector together with the changing needs of manufacturing and trading companies, have all contributed towards the need for warehouse accommodation over the last 20 years. In this respect the investing institutions have responded to the need and provided the UK with a stock of investment warehouse property.

The need for such property has been greatest near centres of population but located close to the motorway network. Developments on or near the motorways coming into London have proved popular with tenants. The construction of the M25 has also changed demand patterns for storage space. This road roughly follows the route of the Green Belt around London. It is interesting to note that the policy of keeping a Green Belt has provided London with a route for a motorway which, as a result, has required very little urban demolition.

Now, however, there is pressure to build on that very same Green Belt. Some of the highest rents for warehouse/low-tech industrial buildings are now being achieved on sites located on or next to this land. In the same way that road building in the south-east in the 1930s caused ribbon development, the motorway system in the 1980s in the south-east is radically changing land use patterns, albeit with the restraining hand of the town planning system.

As previously discussed investors tend to purchase well located industrial buildings which have planning permission for either Class B1 or Class B8 use. It should be realized that speculatively built industrial buildings are only able to accommodate the simpler activities such as light industry or general warehousing. More sophisticated warehousing and industrial techniques require specialist buildings. However, because there are fewer tenants and very little comparable market rental evidence for specialist buildings, institutional investors do not tend to invest in this type of property.

As Figure 5.1 shows, between 1980 and 1986 very little rental growth was

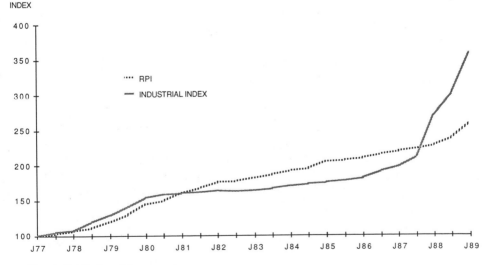

Figure 5.1 Industrial rent index.
Source: Healey & Baker Research.

Figure 5.2 The Mercury Centre, Heathrow, completed in 1988 provides modern industrial warehouse accommodation.

apparent from traditional industrial buildings. During this period, in the southern half of the country in particular, sites which had been allocated for industrial use were developed for foodstores, retail warehousing and housing development.

In the following section we look at the standard specification of a traditional flexible light industrial building. Industrial property investments often consist of industrial estates and this enables a range of buildings to be let, in one location. Management is also easier and a service charge arrangements often exists covering the whole estate.

5.5 A standard traditional warehouse – industrial specification

Whilst there is no absolute specification, investors, over the last decade, developed a general specification for industrial buildings. The aim has been to produce or invest in buildings which appeal to a large letting market requiring a building for warehouse and/or light industrial purposes. A typical scheme is shown in Figure 5.3.

Site cover

There is no ideal site cover (i.e. the area of land covered by buildings). Figures of between 30% and 50% are often acceptable. What is more

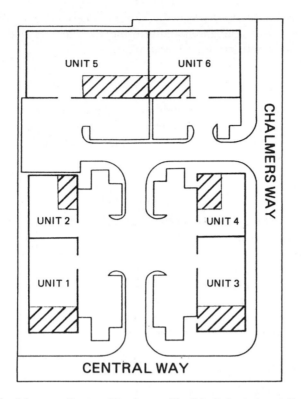

Figure 5.3 The Mercury Centre, Heathrow. Flexible light industrial/warehouse in units from 5755 sq.ft up to 14 608 sq.ft including approximately 10% office accommodation.

important is the way other facilities are provided on the area of the site not covered by the building. The building is normally measured on a gross internal area basis, including common areas.

Layout

A rectangle is normally the most popular shape from a tenant's point of view and also the cheapest to construct. The floor space should be free from vertical stanchions to provide for maximum flexibility. The width between bays varies. For small nursery units of 1250 sq.ft (115 sq.m) the bay may only be 40 ft (12 m) wide. The standard width is in the region of 60 ft (18 m).

Frame

The building is normally a steel portal or sometimes concrete frame providing a clear internal eaves height up to 22 ft or more (6.7 m). This enables a variety of materials to be stored mostly on a multi-level pallet boards system. Fire regulations require that the frame is clad in protective material or faced with protective brick work.

Roof

This is often corrugated fibre cement or plastic coated corrugated steel sheet pitched at about 6° with 10–15% of the roof made of translucent sheeting. On the inside of the roof, panels are installed to provide thermal insulation to comply with statutory regulations.

Cladding

The outside of the steel frame often has brickwork up to about 7 ft (2 m) above which there is PVC-coated metal cladding together with internal thermal insulating panels. The brickwork is more resilient to wear and tear from man, machines and vehicles. A recent trend has been the use of flow cladding in which the roof material flows over the eaves on to the walls. No guttering is provided at eaves height but rain water is collected almost at ground level. This can be a cheaper method of construction but the cladding is liable to get damaged at lower levels.

Floor loading

As the eaves height has increased the demands on the strength of the floor have also increased. The floor should be concrete overlaid with a cement screed sufficiently strong to take fork-lift vehicles and capable of accepting 750 lb/sq.ft (36 kN/sq.m). Loading standards vary but a figure of less than 500 lb/sq.ft (24 kN/sq.m) is likely to reduce the building's lettability and/or cause problems during its use.

Offices and toilets

Only about 10–20% of the building needs to be designed as ancillary office accommodation, which is normally provided at the front. In recent years, as previously discussed, the trend has been to design a layout which permits the office content to be extended up to 50% of the total floor space. The offices should be provided with a vinyl floor covering and toilets are sometimes provided for both the office and the warehouse industrial areas.

Heating

A low pressure water radiator gas-fired central heating system is needed for the office only. Tenants sometimes install fan-assisted heaters or other heating systems in the warehouse area.

Lighting

Sometimes lighting to 400 or 500 lux is provided in the office areas. However, lighting in the warehouse area, if needed, is provided by the tenant.

Services

Gas, water, telephone and 3-phase electricity are provided to the building to enable the tenant to make use of the services as he wishes.

Forecourt and car parking

Externally, access to the building is critical, particularly for large articulated lorries to manoeuvre. The access from the public highway should be unhindered. In front of the warehouse unit there should be a concrete area for the lorries to unload and park off the roadway and this should be at least 50 ft (15 m) deep. The most popular unloading method is by use of fork-lift trucks. A loading dock is an added advantage but not an essential requirement. Within the curtilage of the site, car parking facilities based on a ratio of 1 space to every 500 sq.ft (45 sq.m) of floorspace should be provided.

Loading door

A loading door should be provided sufficiently large to enable a lorry to enter the building. For climatic and security reasons it is not uncommon for lorries to be unloaded inside the premises. It is normally satisfactory for the entrance to the office area to be located on the same side and near the goods entrance.

Externally

The external appearance needs to be clean and tidy with minimal landscaping to keep the maintenance costs to a minimum. If the building is part of an estate, a service charge may often exist to cover the landlord's expenses in managing the estate. This may include providing an estate board stating the names of the tenants, regularly sweeping the estate roadways which should be built to a public highway standard and maintaining the grass and other landscaped areas.

The essence of a warehouse/industrial investment is to provide a building (often referred to as a 'shed' or a 'box') which is well located, will easily let, will produce good rental growth and will not suffer undue functional or material obsolescence.

Tenure

For buildings containing more than about 2000 sq.ft gross internal area (in other words not a seedbed/nursery unit) the market tendency is still to grant 25 year leases with the ability to review the rent paid to the full open market value every five years. Leases are normally on the basis that the tenant is fully responsible for the maintenance, repair and insurance of the structure.

In the case of smaller nursery units, small seedbed companies often prefer shorter leases perhaps of only three years which are outside the provisions of security of tenure of the Landlord and Tenant Act 1954. Such leases deter private investors who seek security of income, hence the involvement of public sector finance for the majority of this type of development particularly as many of these units are let to insubstantial companies who are liable to go into liquidation and be unable to pay the rent.

Whatever the lease structure, where a number of units are developed on one estate, it is essential that an effective service charge account is managed so that the common areas of the estate are well managed to the mutual benefit of the investor and the occupiers. With estates of smaller units, such management may be considerable and detract from the investment value.

References

Bone, R. (1987) Warehousing. *The Planner,* September.

Chiddick, D. and Millington, A. (1984) *Land Management: New Directions,* E. & F.N. Spon, London.

Edward Erdman (1987) *Industrial Space,* Edward Erdman, June.

Healey & Baker (1986) *The Workplace Revolution,* Healey & Baker.

Healey & Baker. *PRIME (Property Rent Indices and Market Editorial),* Healey & Baker (various dates).

McIntosh, A.P.J. and Sykes, S.G. (1985) *A Guide to Institutional Property Investment,* Macmillan, London.

Various (1987) Development economics factories/warehouses. *Architects Journal,* 7, 14 and 21 October.

6

Business use space

Until 1987 the town planning concept of an office building fell within Class II of the Town and Country Planning (Use Classes) Order 1972. Light industrial buildings, as described in the previous chapter, fell within Class III of the same order. The Town and Country Planning (Use Classes) Order 1987 rearranged the uses. An office for the use of financial services and professional companies now falls within Class A2 when used by visiting members of the public whilst a light industrial building and most other types of office or quasi office/business space activity such as research and development, now fall within Class B1.

6.1 The classification of manufacturing buildings

Three distinct although not mutually exclusive types of industrial building can be identified. The first type is a large manufacturing building. These can be over 100 000 sq.ft (9 250 sq.m) and are often located in the areas of the UK which receive most government financial assistance. These buildings employ skilled and semi-skilled labour but may be involved with fairly sophisticated manufacturing processes. Places such as Warrington near Manchester and Silicon Glen near Edinburgh (borrowing the idea of Silicon Valley in California, USA) are examples of such locations.

The second type of building is a quasi-industrial building involved in one or more activities. These may include manufacturing, assembly, servicing, office, retail, storage or promotional use. These service type buildings again tend to be located in south-east England but for two principal reasons. Once again they tend to employ significant numbers of graduates but also they service customers who use their products. These customers in the office and retail sectors are predominantly located in south-east England although there are inevitable smaller sub-centres in all the major towns of the UK.

In recent years this type of service multi-use industrial building has fallen into the category of mid-tech discussed in the previous chapter. Whilst these service industrial buildings may be anywhere from 5000 sq.ft up to 100000 sq.ft or more there are a large number of less than 25000 sq.ft (2300 sq.m).

The third type of industrial building is principally involved with research and development and complex manufacturing for both government projects and private industry. These are primarily located in south-east England. A number of surveys have found that the majority of these employees are graduates, and graduates have a strong preference for living in the southern rather than northern half of the country. These buildings may vary in size from 5000 sq.ft up to more than 100000 sq.ft (460–9300 sq.m). This type of building falls into the category of high-tech or business space mentioned in Chapter 5.

When companies have been interviewed to find out where they would like to locate their highly qualified staff in the electronics and computer industry, they often refer to the concept of 'image'. Image of address and image of building are increasingly important. This was one of the concepts which also emerged from a survey of the office market undertaken by Healey & Baker in 1985 and 1986. It is one of the concepts that government financial assistance to UK industry has only recently begun to appreciate.

Simply pouring public money into locations with assisted area status or into inner city locations is insufficient. Whilst such funds may lower the cost of construction and, over the short term, lower development risk, unless there is an attempt to change the image of run-down areas there will not be long-term sustained demand. As soon as public assistance is removed the area is likely to return to its previous undesirable state unless sustained demand is created.

6.2 Out of town offices

The traditional office buildings will continue to have a preference for town centre locations. The central business district, which can be found in most large towns and is epitomized by the City of London, will continue to attract the major financial organizations. Over the last two decades there have been attempts to split such companies into city centre uses and 'back room' administration functions. Such a trend has not always been successful as 'two cultures' have developed within the same company creating unnecessary friction and managerial problems. There have recently been trends to prevent such dual location companies.

Despite the reversal of a trend to decentralize 'backroom' activities outside Central London there is a growing market for out of town 'campus' style office buildings. These fall within the design criteria for a full office

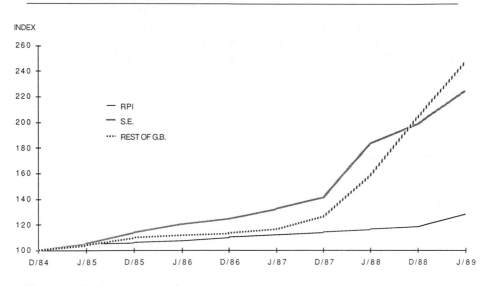

Figure 6.1 Business use rent indices.
Source: Healey & Baker Research.

building although there may be an element of 'alternative' use within the building; part of the premises may be used for storage or industrial activity such as an in-house printing press. The concept of campus offices is subtly but significantly different from the idea of Business Use Space even though both concepts fall within the new Class B1 and much of the business space accommodation created will be for quasi office functions.

In recent years both traditional offices and business space buildings have shown considerable rental growth in the South of England. The growth of research and development activities over the last two decades has not comfortably fallen within a pure office use or that of an industrial activity. The new Class B1 category now does accommodate such use and clear patterns of rental growth are now visible.

6.3 Defining science/research parks

Before discussing the design specification of such a business building it is necessary to clarify the jargon used in the present market.

The concept of a Science or Research Park is not dissimilar to that of a Business Park. The main difference is that there is often a strong link with an academic institution. In many cases the development of such academic parks is aimed at fostering closer links between industry and the academic world. The fact is that in many cases such development in the early stages can not be developed with private funding alone. Public sources, including university funds, need to be called upon often to work in partnership with private monies for instance to provide bank loan guarantees.

Science Park units are specifically designed to appeal to small start-up companies but in design are not dissimilar to small office suites of up to 1000 sq.ft per unit. They are significantly different to low-tech/nursery units. There is not normally a concrete forecourt or a roller shutter door. Instead the building is normally two storeys throughout with central heating but does give the ability to take scientific equipment into the building.

A variation is to provide a fully serviced science based unit where perhaps 20 units within one building are provided. The key to such units is that other secretarial on-site facilities can be provided. Whilst the turnover of the units may be high, some tenants will graduate up, as they grow, into the larger units.

Within the serviced suite building it is also possible to provide other facilities such as a creche, bar, restaurant and even hotel accommodation which may be used by all the tenants on the estate.

There is no doubt that science parks are popular with small but expanding science based companies in the expanding bio-medical and information technology sectors of the economy. The feed back to the local authority or university which carries part of the risk in terms of guaranteeing the income flow back to the fund/investor, is that such development encourages the sponsorship of scientific activity within the academic institution. In an age when much academic research is based on all or part private finance, such an arrangement is to the benefit of the educational establishment.

The most important study into this relationship is *The Cambridge Phenomenon* by Segal, Quince and Partners 1985 which identifies the academic/science based growth of companies. The following were identified as important:

1. There was an interaction between a large number of modern industries;
2. There was a large number of small young independent companies;
3. The area had a record going back 20 years of such development which mushroomed in the 1980s;
4. Research and product design were very important to many companies;
5. The volume of actual production was very low;
6. There were direct and indirect links into the university;
7. More than 70% lived in the immediate area before they started up their new company.

The study by A.V. Bruno and A.C. Cooper, *Pattern of development and acquisition for Silicon Valley Startups*, found that in 1980, of those companies which were twenty years old 37% had ceased trading, 31% were still small independent companies and 32% had been acquired by larger companies.

Of the 50 or more science and research parks which have developed over the last decade more than 30 are linked to an academic institution. The parks which have taken a lead with this type of development are at Herriott Watt in Scotland and the science park at Cambridge. The Genesis Centre at Warrington is a non-university based development started in the 1970s to develop the link between science and industry. From these parks, created in the 1970s, grew a concept which has been slowly modified. As it has evolved other locations have taken on the idea but so far there is no co-ordinated national approach to such development.

6.4 Trade parks

A further title is that of 'Trade Parks' such as the Cheltenham trade park. Like other developments in this area of the property market the idea has been adopted from overseas. It is based on the concept of a mixed-use building, part industrial, part business office space and part showroom – normally to the wholesale trade. Over the next decade such parks are likely to crop up following the concept of 'merchandizing marts' which are found on mainland Europe and in North America.

6.5 Examples of business parks

In recent years a number of landscaped out of town low-rise business parks have developed, the specifications of which are discussed later. Aztec West was in many ways five years ahead of its time in developing this concept. Developed next to the M5 motorway outside Bristol, the Electricity Supply Nominees attempted to create an environment for new high-tech companies in the early 1980s. As can happen with such innovative developments, the design of the units constructed did not meet the needs of the market at that time although the location of the site soon became popular with companies like the Digital Equipment Company taking land for their own development.

Solent Business Park developed out of a planning concept. When IBM were seeking to expand their UK operation in the late 1970s they had diffi-culty finding a site for such development in the southern half of Great Britain. The office buildings available at the time did not meet with their requirements. The result was their own development at North Harbour Portsmouth which required reclaiming land from the sea adjacent to the M27.

This resulted in the structure plan for South Hampshire allocating land for such development in the future. The site of the Solent Business Park provides 150 acres capable of 2 million sq.ft of development as part of a mixed land use of 2000 acres. The success of the idea was vindicated when in 1987 Arlington Securities Ltd sold the first phase to Digital Equipment Co. (DEC) of 94000 sq.ft together with a pre-sale of 66000 sq.ft and 30 acres.

Over the last decade, with the exception of places like the Winnersh Triangle near Reading being developed by Slough Estates, very few large sites similar to developments in North America have become available in the

Figure 6.2 Stockley Park, Heathrow.

private sector. What has happened has been the growth of small business parks and single business units, particularly in south-east England, in an arc following the M25 clockwise from Leatherhead in the south-west including Guildford, Bracknell, Basingstoke and Reading and out along the motorways to the west of London. Stockley Park near Heathrow Airport, created on a former rubbish tip, is arguably the flagship for the larger US style of this type of development. The proximity of the airport reflects the international nature of many but not all of the occupiers of the new business units.

Outside this growth area business unit development is limited to places like the M25 Hampshire, Portsmouth – Southampton corridor, Cambridge, Oxford, Bristol, Warrington/Manchester and West Edinburgh. With the exception of places like Kembrey Park, Swindon funded by Sun Alliance, most of the business parks have so far been developed by property companies, not investing institutions.

6.6 Business use space – a guide to the specification

As with more traditional warehouse/industrial buildings, there is no absolute specification. Indeed one of the concepts behind the development of business units is the flexibility of the financial and design package. This ranges from providing a serviced site for the occupier to design, fund and build upon, through to fully serviced small business suites for new companies where the furniture and secretarial facilities are provided for the occupiers. The emphasis is not on cost-efficient, practical design typical of low- and mid-tech buildings but on the concept of added value. The quality of design must provide a high grade working environment.

The following provides a guide to the most common type of business units developed in recent years.

Site layout and car parking

It has been suggested that no business park is complete without a lake and ducks. Whilst this is not totally true, it does epitomize the high degree of attention paid to creating the right environment which will attract the right tenants and for them to attract the right employees.

It is not uncommon for new development to start with the landscaping by planting semi-mature trees and laying lawns. The buildings then follow. Image of address is important but image of the environment within which the buildings sit is the second leg of the successful double. In some cases

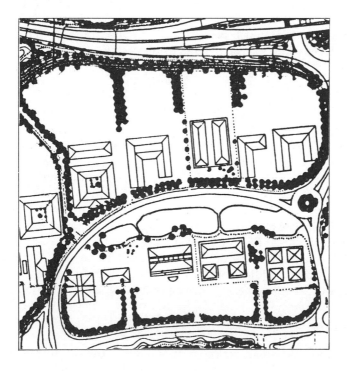

Figure 6.3 Stockley Park, Heathrow. A business space development in a landscaped park.

such as the University of Surrey science park at Guildford, although a science park not a business park, the density of development can be as low as 15%. The density of Stockley Park near Heathrow is similarly very low.

Ideally the car parking ratio needs to be one space per employee. That means that city centres can rarely accommodate such development, as one space for every 250 sq.ft gross is needed and access to a good road network, ideally a motorway, is essential.

Although the site coverage may vary from 30% to 50%, the image of the site is all important. In many cases cars can be accommodated adjacent to the buildings. However at places like Bracknell for instance, a number of developments have been undertaken where the buildings have been constructed above the car parking areas to prevent cars littering the valuable landscape. The key to development is to balance the higher cost of this type of development against the land cost. Where land values are lower separate surface car parking is cheaper to provide and a lower plot ratio and site coverage is applicable. A floor area/plot ratio may vary from 0.5:1 (which is still higher than for many traditional industrial developments) up to 1:1 for the more intensive developments where land costs are higher.

In addition to car access, it is important to remember the industrial nature of these quasi office buildings. Road access for large articulated lorries is important and a site which permits manoeuvring for such vehicles with a 33 ft high access door in each building is necessary.

Construction

Most business units are of ground and first floor only. Some, due to site constraints may have car parking at ground level plus three floors above. Such low rise metal frame construction buildings are generally cheaper to build than conventional office buildings.

The ground floor for large buildings (*not* science park units or small office suites) is often left as shell and core for the tenant to fit out with a 3.5 m (12 ft) floor to clear beams height for flexibility of use. In some markets it may be appropriate to leave the whole buildings in an unfurnished shell and core state for a raised floor and a suspended ceiling to be added later. Each tenant has their own idea of the ideal specification and therefore some developers offer a fitting out package to incoming tenants. The clear eventual floor to ceiling height must be more than 2.7 m (8 ft 6 in).

Roof

Although there are still some flat roof buildings built, in the past both their reliability and appearance have not met with approval, hence the trend towards pavilion pitched tile roofs. This provides the added advantage of preventing solar gain in the summer months.

Cladding, insulation and external appearance

The external appearance projects the image of the occupier which is sometimes important. Full glazing as at Waterside Park, Bracknell and Guildford Business Park developed by London and Edinburgh Trust is important to create an appropriate image to attract occupiers and solar reflecting glass is essential in such situations. In other situations more modest brick cladding is equally successful. As with modern office buildings, but unlike many traditional industrial buildings, insulation both against solar gain and heat loss is important to create a pleasant working environment internally.

The reception area is the most important external feature of the building and it is important that it is made to look imposing to create the right impression for those arriving at the building.

Heating and air conditioning

As one would expect of quasi office space, particularly on the upper floors, an efficient heating system is essential. Due to the international nature of many tenants air conditioning may be essential to achieve the best rental value. The ground floor is often left with no heating equipment but the services are provided for the occupier to connect their own installed equipment into the shell and core accommodation. If an air controlled specialized computer suite is installed such arrangements obviously are most appropriate. The building should be capable of taking either variable air volume heating systems or forced air ventilation. With smaller science park units a simple hot water radiator central heating system will normally be sufficient.

Floors and ducting

As the ground floor is designed for industrial activities, a loading capacity of 500 lb per sq.ft is necessary. The first and other floors, which are intended for office type activities, only require a 100 lb per sq.ft loading capacity. A raised floor is essential on the upper floor of larger buildings but for smaller suites perimeter ducts are usually sufficient.

The upper floor needs to be carpeted with carpet tiles so that modifications can be made and partitioning installed easily. It is debatable whether carpeting is necessary on the ground floor. Sometimes heavy duty industrial carpet is appropriate for certain markets. Although the ground floor will often be fitted out by the tenant to his own requirement, at first floor level adequate ducting perhaps by way of a raised floor is the key to efficient cable management.

Lighting and ceiling

Good natural daylight is necessary for the upper floor(s) plus artificial lighting to provide 500 lux within a fully integrated suspended ceiling. If the depth of the building does not give good natural light from the side windows, roof lighting may be necessary. On the ground floor it is normally appropriate for the tenant to install the necessary lighting.

Kitchens and toilet

The quality of toilet facilities creates a strong impression on potential tenants. On both ground and upper floors they need to be adequately provided to a high standard of finish. Kitchen facilities in large buildings are also necessary although in smaller buildings of, say, less than 30000 sq.ft it is only necessary that the services are installed for the facilities to be added at a later date if necessary.

Services

It stands to reason that all main services should be provided including a three phase electricity supply. Due to the high-tech nature of modern tenants the first floor must be on a raised floor system with adequate power for a visual display unit (VDU) and other electronic equipment to be located at each work station.

Lift

All business units, if they are to attract the best tenants and the best rental values, must provide at least an eight-person passenger lift. In larger buildings it is also necessary to provide a goods lift.

Tenure

As previously mentioned the key to success for business parks is the flexibility of both the design envelope and the financial/tenure package offered. The traditional lease used by institutional investors may be inappropriate if the tenant is unable to maximize his benefit from plant and machinery taxation allowances.

There are three methods of arranging tenure. First, particularly for new small companies, a long lease may be inappropriate as it is well known that some small companies soon grow into large concerns demanding more accommodation. For them a short lease, perhaps for only three years, is appropriate outside the provisions of security of tenure of the Landlord and Tenant Act 1954. For many private investors such tenure does not give them the security of income flow which they would like; hence the active involvement of public funds for seedbed, start-up business units.

The institutional 25-year lease with five-yearly rent reviews to the then open market value of the premises is more appropriate for private funded investors. For larger public quoted companies as tenants this rarely poses a problem although for small companies where director guarantees to the lease are required, such long-term personal financial commitments are disliked. For American companies, where a commitment to pay rent over time is treated as an accountancy liability, such an undertaking is also not prefered. A short five or ten year lease is sometimes negotiated by such tenants.

The third method of tenure is to sell a qualifying interest for tax purposes. This may mean actually selling part of the freehold of a business park or a long (125 year) leasehold interest at a peppercorn rent in consideration for a premium payment.

Leasing finance as discussed in Chapter 4 may be one route where a financial organization in effect purchases a qualifying interest for tax purposes and then charges the tenant interest plus repayments over a period of time, perhaps 25 years. Due to the nature of the loan, Plant and Machinery Allowance on a 25% depreciating balance basis and an Industrial Building Allowance on a 4% straight line basis, will be claimed by the finance house and preferential lease finance terms shared with the occupying tenant.

The more specialist the type of building that is constructed on a particular site, the more likely the tenant is to purchase the freehold or long leasehold. That owner/occupier may then set up their own taxation allowance arrangements. The disadvantage of this arrangement is that the occupier may design and construct a specialist building which will have to be written down at cost and will have little if any market investment value as time progresses. Depending on the location, in a very few years (perhaps less than five) the site may become more valuable than the building upon it!

The key to the long-term success of business parks is not only a well designed infrastructure, such as the road layout and the landscaping but the continual maintenance of the park. For this reason, whatever method of tenure employed by occupying tenants, and there might be a mixture of methods used within a particular park, all tenants/occupiers should contribute towards a service charge for the upkeep of the estate.

References

Carter, N. and Watts, C. (1984) *The Cambridge Science Park, Planning and Development. Case Study No. 4*, RICS, London.

Coopers and Lybrand Associates Drivers Jonas (1986) *Accommodation Needs of Modern Industry.*

Debenham, Tewson and Chinnocks (1983) *High Tech: Myths & Realities.*

Fuller Peiser (1987–88) *High Technology – B1.*

Hall, P. and Markusen, A. (eds) (1985) *Silicon Landscapes*, Allen and Unwin, London.

Healey & Baker (1986) *The Workplace Revolution*, Healey and Baker.

Healey & Baker *PRIME (Property Rent Indices and Market Editorial)*, Healey & Baker (various dates).

Henneberry, J. (1987) *British Science Parks and High Technology Developments: Progress and Change 1983–1986*, PAVIC Publications, Sheffield City Polytechnic.

Herring, Son and Daw (1982): *Property and Technology – The Needs of Modern Industry 1982*.

HMSO: *Town & Country Planning (Use Classes) Order 1987*. HMSO, London.

Monk, C, *et al.* (1988) *Science Parks and the Growth of High Technology Firms*, Croom Helm (in association with Peat Marwick McLintock), London.

National Development Control Forum (1985) *High Technology Development The Planning Control Considerations.*

Various (1987) Development economics, offices/industrial buildings. *Architects Journal*, 20, 27 May and 3 June.

Waldy, E.B. (1986) *Business Parks*, Fletcher King.

Williams, J. (1982) *A Review of Science Parks and High Technology*, Drivers Jonas.

7

Workspace and business centres

7.1 Introduction

Workspace developments are schemes involving the creation of accommodation for small businesses in the form of separate or communal units, through the conversion of existing buildings or new construction.

The other characteristics of both workspace developments and business centres have been an active on-site management, along with the provision of common services and flexible letting terms. Within the broad definition of workspace are a series of sub-categories including managed workshops, seed-bed centres, incubators, business centres and technology centres.

Workspace developments are the response to a number of problems and opportunities. Research and experience throughout the UK has shown frequently a shortage of suitable accommodation for new and very small established businesses. Furthermore, workspace projects are thought to promote enterprise development by providing supportive environment for small businesses generally, or specifically for some sectors such as new technology firms. They have also been undertaken as a special measure in areas of rapid decline in traditional industries, such as in steel closure areas, aimed at replacing lost jobs. Other initiatives in this field have been motivated by a desire to reuse old and redundant buildings.

Although workspace developments are the response to diverse problems and opportunities, they do not represent a single response. There is a great variety of different types of development. The simplest classification is based on the distinction between conversion and new-build schemes and on the inclusion or otherwise of common services. In practice, there is a continuum of different packages of unit types and services, involving various forms of business development support and, often, subsidies. The creation of workspace developments has relied to a considerable extent on public authorities and a number of development trusts and agencies. The role of developers and other private organizations is restricted to certain types of development, e.g. small factory estates and simple subdivision of easy-to-convert industrial buildings mainly occurred due to the tax incentives provided by the IBA scheme between 1980 and 1985 (Chapter 11). The involvement of financial institutions is very limited. Development and funding arrangements are often complex and unconventional, and the development process can be very slow, especially in the case of conversion schemes.

In recent years some developments have become more specialized, aiming at particular types of companies. The most common specialized workspace development is the business centre, aimed at office-based businesses, often requiring a higher standard of accommodation and services. Another specialized development is the technology or innovation centre, focusing on small firms involved in hi-tech industries. Both of these types of development will be examined later.

Background

Workspace has its roots in the pioneering development championed by architect David Rock at 5 Dryden Street in Covent Garden, London, in 1972. The scheme involved the conversion of a multi-storey Edwardian building into 17 000 square feet of open plan office space. The novel features (at the time) were the extensive provision of common services, and the involvement of the tenants in the management of the development.

Hope Sufferance Wharf in Rotherhithe, London, was another early example of a workspace development, converting a historic development into craft workshops in 1973. Perhaps the catalyst for a number of developments, however, was the Clyde Workshops development undertaken by BSC (Industry) in Glasgow. Although unremarkable by the standards of subsequent developments, Clyde Workshops attracted considerable attention at a time (1979) when the issue of premises for small firms was becoming topical.

There is no official estimate of the number of workspace schemes in the UK. A rough estimate would suggest some 500 schemes spread throughout the country. Their geographical distribution depends on how active the public authorities are, or have been in a particular locality. Thus, Urban Programme areas, where funds have been available for developments, have a number of schemes. Similarly, areas such as Strathclyde, also have a number of schemes, partly due to the Regional Council's commitment to new enterprise workshops. On the whole, those areas in receipt of Urban Programme funds and those with local authorities active in economic development have a number of developments.

Although the initial enthusiasm for workspace development has abated, there appears to be no reduction in the number of new schemes being developed. What has happened is that public and private sector developers now have a clearer idea of the costs involved, and the requirements for a successful development. There is now a more business-like approach to new development, rather than subdivision of a building, merely because it was available.

7.2 Types and characteristics

The large number of workspace developments and major differences in their objectives and approaches, from design to management matters, make it very difficult to devise a satisfactory classification system. An initial attempt to categorize schemes is by the physical characteristics of the development.

Physical characteristics

NEW-BUILD SCHEMES WITHOUT COMMON SERVICES
The first type is that of a modern single-storey unit industrial estate without on-site provision of shared or other support services. Developments of this type are normally built to a high standard of construction and vehicular circulation. The factory units are often small, below 1250 sq. ft, but very rarely below 250 sq. ft. There are a great many examples throughout the country. Kingston upon Hull City Council has developed seven small unit factory estates comprising a total of 98 units. Only 10% of the units are larger than 2500 sq. ft; 27% are between 1250 and 2500 sq. ft; and the majority (63%) are below 1250 sq. ft, the smallest being 320 sq. ft.

NEW-BUILD SCHEMES WITH COMMON SERVICES
The second, and much less common, type involves a modern small factory estate as above but with the provision of shared services. This is the case of the Innage Park development at Atherstone, North Warwickshire. The scheme involves 17 units of 600 to 1000 sq. ft. It was developed through a lease-and-leaseback arrangement between North Warwickshire District Council and Audit and General Dvelopments Ltd, who subsequently sold their interest to a pension fund. The Atherstone Factory Association was formed to manage the Innage Park Estate and to organize the provision of shared services, including typing, telex, telephone answering, accounting and canteen.

CONVERSION SCHEMES WITHOUT COMMON SERVICES
The third type involves conversion schemes creating separate small units without common services. An example of this type of development is

Imperial Studios in Fulham, London. The 20000 sq.ft scheme was developed by a private company, Local London Group, with a loan from the London Borough of Hammersmith and Fulham. It was subsequently sold to the London Small Business Property Trust. The Hope Sufferance Wharf development is another example of a conversion scheme without common services, whilst there are some cases, such as the Templeton Business Centre in Glasgow, which provide very limited services and cannot be classed as common services projects. Templeton only offers to firms access to conference room facilities. Tenants have also the opportunity to obtain business advice from the office which the Scottish Development Agency has on site. In this category also belong projects where businesses have subdivided their surplus floorspace and a large number of entirely private sector conversion schemes.

CONVERSION SCHEMES WITH COMMON SERVICES
Shared services represent a key characteristic of many workspace developments. Two cases representative of the fourth type, involving conversion to separate units plus shared services, are Mantra House in Keighley and the Port Talbot Workshops. Mantra House is a former engineering works comprising two storeys which has been converted to 67 units by the City of Bradford Metropolitan Council. The development provides telephone answering, typing, photocopying, foreign translation and other facilities. Similar facilities are also provided at the Port Talbot Workshops, a 48 unit scheme created by BSC (Industry) Ltd through the conversion of a modern single-story unit.

A subcategory of this type of scheme is that based on an open plan layout, where normally each firm's workspace is delineated with fixed or temporary partitions. 5 Dryden Street is such a scheme. Another example is the Avondale Workshops in Kingswood, Bristol. The scheme was created by the New Work Trust Company Limited and provides 66 workspaces together with extensive support services.

Types of enterprise

At another level it is possible to categorize developments by the type of companies they seek to attract, or the type of support services they are designed to offer. The main types are:

MANAGED WORKSHOPS
A general term for a development offering basic space normally to both service or manufacturing firms, with relatively few restrictions. Selection of tenants primarily based on simple criteria, such as ability to pay. Examples include a large group of schemes developed by the GLC and now run by London Industrial PLC.

SEEDBED WORKSHOPS
Combine small amounts of space and common services with subsidized rents to assist new businesses in the start-up period. Barnsley Enterprise Centre was established by Barnsley Metropolitan District Council using Urban Programme money. The rental services charges are introduced gradually with the first three months free. Tenants are normally restricted to a maximum stay of 12 months.

NEW ENTERPRISE WORKSHOPS
Closely related to seedbed workshops, with additional emphasis placed on the provision of communal equipment, and the back-up of staff with technical expertise. An example is the Community Enterprise Centre in Birmingham.

BUSINESS CENTRES
Developments designed to attract office-based business, normally offering a higher standard of accommodation and specialized common services (e.g. fax, computing), e.g. the chain of business centres established by Local London Group PLC in London.

INNOVATION/TECHNOLOGY CENTRES
Developments established to promote new technology and innovation in an area. These developments tend to be highly selective in the choice of tenants. Recent developments include the Business and Innovation Centres in Barnsley, Cardiff, Lancashire (Rawtenstall) and Clwyd (Newtech).

The above list is not exhaustive. There are, for example, a number of workspace developments aimed at community businesses or ethnic minorities. The five categories do however encompass the majority of workspace schemes and can be collectively described as 'managed workspaces'. In simple terms the developments either have a mission which determines the types of client they aim for and the services they provide, or they are targeted at a particular gap in the property market.

Managed workspaces

These developments on the whole share four common characteristics.

(i)   Size of space. Most workspace developments offer a range of space from the very small (100–200 sq. ft or even less) to around 2500 sq. ft. They fill a gap, particularly at the bottom end of the market, which the private sector had ignored.
(ii) Active management. The majority of workspace developments have an active on-site management. This can range from a receptionist up to a fully staffed office with manager, receptionist, secretary, book-keeper, handyman. The level of staffing is normally determined either by the

size and type of the development or the amount of subsidy available from a public authority. A 30 000 sq. ft facility will generally support a management team of a manager and receptionist, plus part-time book-keeper and secretary. 50 000 sq. ft will support two additional staff – a business adviser and a maintenance technician.

(iii) Flexible letting arrangements. The majority of workspace developments offer very short-term leases or licences, minimizing the commitment of the small businesses. As little as one month's notice is required. In addition, they usually offer an all-in payment incorporating rates, rents and communal charges such as cleaning, etc. In certain developments, heating costs are incorporated in the charge.

(iv) Common services. Almost every workspace development provides some form of common services. These can include property services (heating, lighting, cleaning, security), office support (reception, switchboard, mail, telex, photo-copying, typing), communal space and equipment, and business counselling. The services available are determined initially by the objectives of the development and ultimately by demand. The costs of some of the services are incorporated into the licence or rental fee while others are paid for as used by each tenant.

In terms of physical characteristics the vast majority of managed workspace developments are conversion schemes with common services. New-build schemes with common services are very rarely found outside the London area.

Overall categorization

When one takes into account a wide range of characteristics, e.g. type of tenant firms, type of building, type/extent of common services, basis of rent/charges, and development/funding arrangements, a composite typology emerges. In such a typology one can distinguish between low-cost/overtly subsidized conversion schemes for enterprises in the process of setting up, and conversion schemes operating on a break-even basis and catering for established but young firms. Overall, the wide range of unit sizes, finishes, location of units within multi-storey buildings and the implications in terms of access, etc, as well as the variations in facilities and common services, make workspace developments in conversion schemes a very flexible and varied form of accommodation which is likely to meet the needs of different types of small firms in the manufacturing and service sectors. Conversion projects are particularly suited to very small and new firms, especially when their flexibility is combined with low costs to the tenants.

Large-unit conversion of single storey and other less old buildings tend to fall into the same category as new-build schemes. They are undertaken

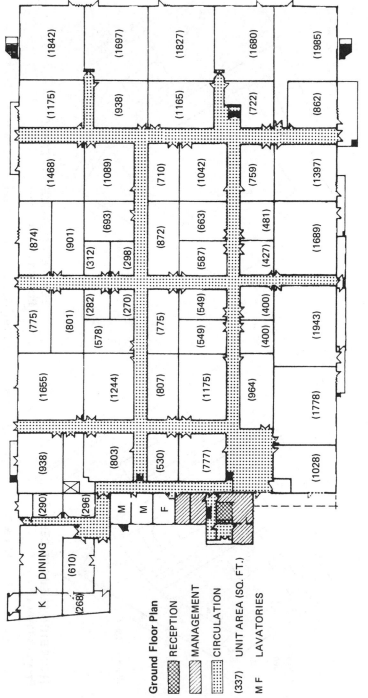

Figure 7.1 Station Road workshops, Bristol. Fitted and fully serviced, self-contained workshops (200–2000 sq.ft).

Ground Floor Plan

RECEPTION

MANAGEMENT

CIRCULATION

(337) UNIT AREA (SQ. FT.)

M F LAVATORIES

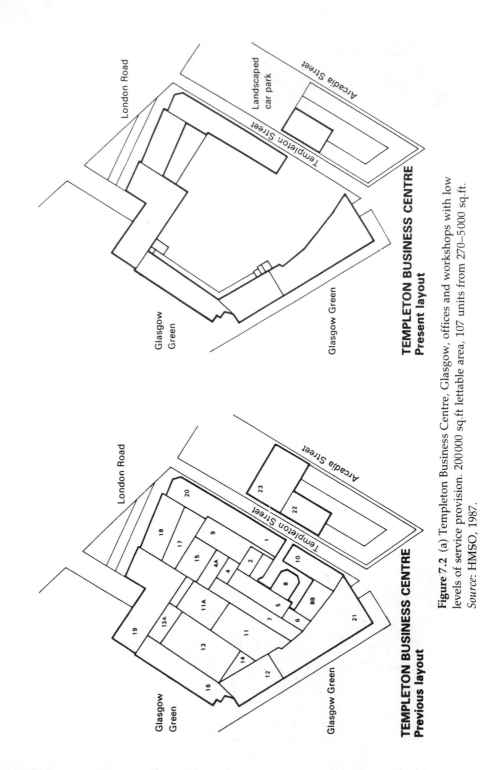

TEMPLETON BUSINESS CENTRE
Present layout

Figure 7.2 (a) Templeton Business Centre, Glasgow, offices and workshops with low levels of service provision. 200000 sq.ft lettable area, 107 units from 270–5000 sq.ft.
Source: HMSO, 1987.

TEMPLETON BUSINESS CENTRE
Previous layout

(b) Templeton Business Centre, Glasgow.

commercially and cater for growing small firms which require their own self-contained premises. There is less variety and flexibility in the small factory estates but these are more suited to the more established and growing small firms which need to project their own identity and can be self-sufficient as far as ordinary business services are concerned. Such businesses require factory or warehouse accommodation with direct access and very limited or no common services.

7.3 Costs

Development costs

In terms of overall size, workspace developments vary widely. However, very few are smaller than 10 000 sq. ft and very large schemes of, say, over 100 000 sq. ft (e.g. Templeton Business Centre) are the exception. The majority are between 20 000 sq. ft and 40 000 sq. ft. This distribution is also reflected in the development costs. A large proportion of developments had a total cost in the region of £500 000 (e.g. Port Talbot Workshops £415 000, Merseyton Road Workshops, Ellesmere Port, £510 000), whilst there have been some low-cost schemes (e.g. Saltaire Workshops, Bradford, £156 000) and some schemes with a high total cost (e.g. Glasgow's Templeton Business Centre £2.5 m, and Govan Workspace £1.3 m).

Although one can identify a mainstream of schemes in terms of total development cost there is an enormous diversity in terms of unit cost. This ranges from well over £20 per sq. ft of lettable floorspace, e.g. Mantra House and Templeton Business Centre, to less than £5 per sq. ft., e.g. Saltaire Workshops and Clyde Workshops. Major variations can exist even within the same project as in the case of Govan Workspace where the unit cost at one of the sites, a former bakery, was twice as high as that of another, a redundant school building (£15 and £7.50 respectively). As a very general rule at present development costs of a 30 000 sq. ft scheme should be below £15 per sq. ft (£20 for office use) if subsidies are to be avoided.

The unit cost in conversion schemes reflects many factors, including the value of the property, the structure of the building (e.g. the need to install lifts in buildings with several storeys) and its condition (e.g. the extent of roof and other major repairs), the availability of main services, especially electricity and gas, and of toilet facilities, etc. It also depends on whether only limited repairs are to be carried out initially with the possible consequence of high maintenance costs thereafter. Another important factor is the desired specification, which is derived from the nature and the objectives of a project and concerns the quality of partitioning, lighting, heating and fittings, whether gas or electricity are to be individually metered, the provision of false ceilings and floor coverings, etc.

As a very general rule development costs currently usually range from £10 to £30 per sq. ft for conversion schemes. For new-build schemes they are in the region of £25 per sq. ft for single storey blocks and over £35 per sq.ft for two-storey ones, excluding land acquisition. For a typical scheme of 30–40 000 sq. ft, which is the minimum size for a financially viable scheme in most instances, the development cost is likely to be in the region of £1m.

Operating costs

The operating costs vary widely depending on the extent of common services and the intensity of on-site management. The running costs of fully managed workspaces will include substantial outlays for staff salaries, office expenses (telephone, stationery, etc.) as well as property service expenses such as cleaning of common areas and building insurance. Security and maintenance/repairs could also be substantial elements of expenditure, and the latter could prove a major liability especially in older buildings.

For a 30–40000 sq. ft managed workspace development this would normally mean an operating budget of £70–100000 p.a. excluding the repayment of development finance which is not always fully reflected in operating budgets due to various forms of subsidy. In practical terms these levels of expenditure would either require substantial rental income from high-rent uses (e.g. computer firms) or some form of subsidy for a scheme to break even or produce a return on capital invested.

Finance and management policy

Rents or licence fees and service charges represent the mainstream of income of workspace developments. Some schemes also generate small amounts of income from services offered to outside firms and from other activities, such as consultancy and training. In managed workspace schemes it is common for service charges to equal, or even exceed, the rent or licence fee. Such a gross rent could therefore be much higher than the net rent in normal commercial property and often becomes a point of friction between management and tenant firms. There is therefore the tendency for management to bill tenants separately for service charges, wherever possible, and to charge on a pay-as-used basis for discretionary services and facilities, such as photocopying, telephone calls, use of meeting rooms.

The overall cost to the tenant firms, viz. inclusive rent or licence fee, depends on the location, the quality of accommodation and services and other factors. As an indication, inclusive rents/licence fees in a wide range of workspace development in Bristol in 1988, were between £3.50 and £7.00 per sq. ft for light industrial workshop units. However, it must be pointed out that some of these charges are fully inclusive, covering rates and even the heating of individual units whilst others exclude a number of cost items. It is now increasingly the practice to exclude rates and telephone from the inclusive rent or licence fee.

Another major factor affecting rent levels and the income of workspace developments is the objectives of the schemes. These may dictate a selective letting policy in favour, generally, of start-up business or, specifically, by unemployed people or other priority groups. This in turn often entails a policy of below-commercial-level rents, at least for an initial period of one to three years. Seedbed workshops and new enterprise workshops normally operate such policies.

In other cases the workspace scheme is run on a commercial basis, which, in practice, means seeking to charge rents similar to those pertaining to comparable accommodation, and to cover the full cost of services and facilities. In reality many schemes are not of sufficient size or quality or in a high-rent location to generate income commensurate with their outgoings and they resort to various forms of subsidy. Local authorities meet such budget deficits in many of the schemes, especially managed ones, and in several cases, receive government support through the Urban Programme. Another way of maintaining a quasi-commercial approach is to write-off part of the development cost but to insist, thereafter, on a non-subsidies approach.

A large majority of workspace developments have been funded with the benefit of public subsidies or other incentives. Managed workspace schemes typically have received Urban Programme grants through the relevant local authorities. Mortgage finance has provided a secondary source of funding for some schemes promoted by independent development trusts.

The absence of a prospect for an attractive return on investment and a dislike for the management complexities of workspace schemes, has meant that private funding has been restricted to a few categories of developments. Private developers have undertaken conventionally funded small-unit industrial estates with the benefit of the now discontinued IBAs and, in some cases, local authority guarantees, too. However, pension funds and other financial institutions are almost totally absent from the managed workspace scene even in the case of business centres. In the latter case the developers tend to rely on their own finance and short- to medium-term bank loans to undertake the development. The same applies to the subdivision of large factories by industrial companies or developers. A different form of private sector participation has been the sponsorship of major managed workspace schemes by large private companies, as part of their social responsibility approach, as in the case of enterprise centres funded by BAT in Liverpool and Brixton.

A new government initiative in 1988, aiming to plug this apparent gap in commercial funding, has been the new role accorded to English Estates by the Department of Trade and Industry. English Estates have now the brief to develop managed workspaces in inner city areas, charging an economic rent but aiming at a relatively low 5% yield, and seeking partnership with the private sector.

7.4 Special types

Incubator workspaces

The term incubator workspace is a general term for workspace developments which aim to provide a supportive environment for small businesses. On the whole these are managed workspace developments which operate at the bottom end of the market, offering small, unpretentious space to small and very small firms.

The principal characteristic of these developments is an active management. This can vary from a caretaker/handyman up to business and technical support for tenants. The majority of incubator workspaces were established with public sector support. The result is that they have objectives beyond merely offering premises, such as helping new businesses survive the first few years of trading. Thus the management attempt to build close relationships with tenants to counsel/advise as necessary, and where appropriate, bring in outside advisers.

On the physical side, incubator workspaces tend to provide small, basic units. This is in part due to the types of building which have been converted, and an attempt to minimize the costs charged to individual tenants. These schemes tend to cater for all types of businesses, service or manufacturing.

With regard to common services, most incubator workspace developments offer telephone answering, photo-copying, typing and book-keeping facilities. the emphasis is on only paying for services used, thus minimizing costs. A number of developments such as new enterprise workshops do offer technical support services.

Business centres

Business centres are effectively up-market managed workspaces aimed at office and some hi-tech businesses. They are a fairly recent development in the UK, a response to the continuing growth of the service sector. Business centres tend to be located in or near existing city centres. One feature which is noticeable is that there are a number of business centres operating without any public sector support, indicating that this part of the market is more profitable than for managed workspace generally.

Business centres, like other managed workspaces, have an active management. Generally speaking there is less hand-holding than with incubator units. This is in part due to the fact that business centres tenants have either been self-employed before, or are familiar with business issues.

The standard of premises offered by business centres is high. The space offered can range from a desk and telephone to a fully furnished suite of offices. The standard of finishing is usually quite high, including carpeting and office furniture. Business centres tend to be selective about their clients, and their charges, on a square foot basis, can be high compared to managed workspaces.

The types of common services offered by business centres vary and are not necessarily more extensive than those offered by incubator workspace developments. Many schemes, however, include telex, fax, and translating facilities in addition to the typing/telephone answering facilities provided as standard. The emphasis in schemes at the top of the workspace market is on encouraging the use of information technology and on providing a service environment up to international business standards.

Innovation and technology centres

Innovation and technology centres provide an environment in which new products and new businesses with a high growth potential can be fostered. The early innovation centres at Hull and Merseyside have been superseded by a network of European Commission supported business and Innovation Centres (BICs).

The emphasis in innovation and technology centres is on a very active management team, with a very close relationship between staff and tenants. The classic type of tenant is one with considerable technical/innovation skills but limited business skills. The centre's main objective is to help the entrepreneur translate the idea into a commercial product.

Accommodation standards in the latest technology centres tend to be high, similar in some ways to business centres. This is particularly true where new information technology businesses are the target. The standard in the older centres, where many new products were in the mechanical engineering area, tend to be similar to the standard managed workspace.

Common services in the older innovation centres related to shared machinery. In the newer centres the emphasis is on consultancy support, often through in-house staff, and access to information sources, such as international databases and sources of venture capital. These are in addition to the fax, telex, etc. facilities provided in business centres.

7.5 The contribution of workspace schemes and business centres

There has been no overall assessment of the success of workspace developments. This is partly due to the diverse nature of developments in terms of size, funding sources, and objectives. There have, however, been a number of studies which have looked at a limited number of schemes. One problem which occurs in every study is to define the word 'success'. Is it success from the tenant's point of view, from the management's point of view or from the funder's point of view? Is it success in meeting the original objectives, in providing premises cheaply, in nurturing new businesses? There is no easy answer. All we can do is assess developments from a number of viewpoints and highlight drawbacks/problems from which others may learn.

Tenants point of view

From the tenants point of view the principal reasons for locating in a workspace are a combination of space availability, reasonable letting terms and low rents. The small businesses' costs are minimized by occupying the small amount of space required in a development which does not demand a long-term commitment.

Evidence does suggest that tenants use of shared services is more limited. Common services are not regarded as vital by small businesses and they do not constitute a prime reason for locating in a development. In most cases many of the services offered (e.g. typing, book-keeping) can be undertaken by friends, relatives, etc. or from an outside agency. This is not the case with business centres or innovation centres where many of the shared services would require what can be substantial investment by the individual business.

The level of demand for shared services is affected by a number of factors including the availability of alternative services in the immediate vicinity, the activity of the firms (shared services are normally more relevant to office businesses) and the age of individual businesses, since new businesses tend not to have secretaries, photocopiers, etc. Many developments have sought to increase their income by marketing their services to companies outside the development.

The other factors which determine the tenant's view are the quality of the manager and access to the development. It is now accepted that the manager has a key role to play, alternating between arbitrator, counsellor, diplomat and law enforcer. Access to the development is an important issue, not only in relation to parking/loading, but also condition of lifts, access outside normal working hours, etc.

Employment generation

Judging workspace developments contribution to economic development is more difficult. There is no doubt that workspace developments have removed constraints in the local property market. Small firms start up and operating costs are reduced by the availability of small units. However, it is impossible to quantify how many jobs/firms are still in existence because of this.

The same reasoning applies to the intangible benefits of workspace developments, such as positive environment, common services. Again, it is impossible to quantify any impact, although it is likely to be greater in those offering a comprehensive package – seedbed centres, new enterprise workshops and innovation centres. The indications obtained from specific studies, e.g. on the performance of managed workspaces in Cleveland, suggest that workspace provision can be an effective form of support to enterprise creation but it does not necessarily have a major impact on enterprise development – the subsequent growth and success of businesses.

Financial viability

The financial performance of workspace developments is another aspect to be considered in assessing their success. Many developments aim to break even or show a return through income exceeding operational costs rather than total development costs. It should be pointed out, however, that although many schemes show a modest return for their sponsors, they would not be viable if all costs were included or if a full commercial rate of return was required on the investment. There is evidence that very often hidden public subsidies of well in excess of £20000 cover the cost of feasibility studies and other preparatory work and that there are other below cost elements which assist the viability of such schemes. There is clear evidence that without such subsidies and without the schemes' sponsors accepting a lower rate of return, many developments would not materialize.

This financial situation together with the disincentive of a very complex and slow process of translating a conversion concept into an operational scheme, tend to rule out the conventional property industry and place the onus on local authorities and other development bodies. Departures from this rule occur when exceptional financial circumstances, such as the availability of a suitable building at a very low price, allow a development company to convert existing buildings to small units. A study by Roger Tym and Partners (1984) has shown that small developers specializing in this field can undertake such schemes in areas of high demand using internal funds, especially if the project involved a low-cost break-up conversion of larger units.

The financial picture is even less encouraging in the case of pay-as-used common services. Recent studies show that they are not yet fully proven, especially the more advanced office and business services. Information available suggests that management-operated services have a shortfall of income in comparison with actual costs. The weak financial performance of workspace schemes, often means that developers cannot afford to secure the services of a high calibre full-time manager, who is the key to success of managed schemes. There are, however, some schemes sponsored by development trusts which have the benefit of such high quality management since their founders have stayed on to manage the schemes, accepting a lower remuneration than they could secure in the commercial sector.

The overall operational costs of workspace, including the cost of management and common services, are high. Their financial viability depends on achieving a high rate of occupancy, often 90% or more, and a fairly high income from service charges, plus various forms of subsidy. For example, the turnover of occupants can be very high, especially in schemes catering for very young firms.

However, a high turnover is inevitable in developments which make it easy for new businesses to locate there and assist them to become established and grow. Nearly 70% of the occupants in Clyde Workshops are new businesses and experience shows that 60% of the firms leaving the scheme do so in order to grow, whilst 40% leave because they fail. Notwithstanding a high turnover of occupants a majority of workspace developments manage to maintain a high occupancy rate with the help of attractive rent levels and especially flexible letting terms.

Letting difficulties, however, are not unknown and serve to emphasize the importance of marketing a development. Where demand is limited or where competition exists between schemes in a given area, success will go to the developer who designs the most effective marketing mix of product (workspace of appropriate size and quality), place (an accessible location), price (rent and other costs) and promotion (in which the workspace manager can play a key role). The promotional package may have to include the offer of a setting-up grant or phased rentals.

7.6 Conclusion

Most workspace developments and business centres would not have developed without the financial support or subsidy of the public sector. It is clear that new-build or conversion costs, combined with the relatively high running costs make these developments unattractive to the majority of private sector developers. The experience of the 1980s has allowed a considerable body of expertise and knowledge to accumulate, and this is likely to lead to more carefully planned and targeted developments.

Many earlier schemes were undertaken when unemployment was at its peak and the pressure on local authorities was to be seen to be doing something. The political pressure, combined with other local factors, often led to developments which, with hindsight, could never be viable. At the same time, claims about the benefits of workspace developments were hard to resist or difficult to disprove.

In spite of these drawbacks workspace developments have proved popular with small businesses and have undoubtedly filled a gap in the property market. The job-creating potential of workspace has now passed from the run of the mill development to the specialist business centres and innovation centres. However, early claims of their success might ultimately lead to disillusionment when their job-creation results are set against their financial performance.

Perhaps a belief that the provision of premises for new small firms should be self-financing is unrealistic. The Enterprise Allowance Scheme is not self-financing, nor is the Department of Trade and Industry's Enterprise Initiative, or Adult Training Grants. The authorities have spent considerable amounts of money subsidizing small business costs through various schemes, so why not through premises. Nevertheless, there is an urgent need to achieve greater clarity in the objectives of workspace schemes especially in the way they are reflected in management policy, and their financial repercussions. In the final analysis it must be possible to establish beyond any doubt whether and how any subsidies are translated into effective support for enterprise and employment creation.

References

Coopers and Lybrand Associates with Driver Jones (1980) *Provision of Small Industrial Premises*, Department of Industry, London.

Department of Industry and Shell UK Ltd (1982) *Helping Small Firms Start Up and Grow: Common Services and Technological Support*, Report prepared by Job Creation Limited, HMSO, London.

Department of Trade and Industry (1985) *The Small Workshop Scheme: A Review of the Impact of the Scheme and an Assessment of the Current Market Position for Small Workshops*, DTI, London.

Haris Martinos (1985) *Workspace Development for Small Business*, The Planning Exchange, Glasgow.

Haris Martinos (1987) *A Guide to Managed Workspace Schemes*, PA Cambridge Economic Consultants/Department of Trade & Industry.

Haris Martinos, *et al.* (1988) *Cleveland Managed Workspace Study*. Report to The Cleveland Co-ordinating Team.

Howard Green, Karan Boyland and Adam Strange (1988) *Management Workspaces in Yorkshire and Humberside: An Analysis of Employment Generation and Tenant Performance*. Working Paper No. 3, Department of Urban Planning, Leeds Polytechnic.

Jackson, Mair and Nabarro (1987) *Managing Workspace*, HMSO, London.

Job Creation Limited and Segal Quince & Associates (1984) *Helping Small Firms Start Up and Grow: The Impact of Managed Workshop Schemes*. Report to DTI and Shell UK Limited.

Paul Foley and Howard Green (1984) *A Practical Guide to the Conversion and Subdivision of Industrial Property*, Howard Green & Associates, Ilkley.

Perry, M. (1986) *Small Factories and Economic Development*, Gower, London.

Roger Tym & Partners (1984) *Mills in the 80's*, Report to Greater Manchester Council and West Yorkshire County Council.

Segal Quince Wicksteed (1985) *Case Studies of Two Managed Workshop Schemes*, HMSO, London.

URBED Ltd (1987) *Reusing Redundant Buildings*, HMSO, London.

Acknowledgements

The help of John McCreadie, until recently with The Planning Exchange and Pascale Eidman-Barnes, IDP Ltd, in preparing this chapter is gratefully acknowledged.

Part Three
IMPLEMENTATION AT LOCAL LEVEL

8

Local economic and employment development

8.1 Introduction

Local economic and employment development is a relatively new concept, having evolved over the past ten or twelve years as a response to a number of diverse factors. At its simplest, it can be defined as measures/policies aimed at encouraging economic and employment development at the local level. Although defining the word 'local' can be a problem, a wide definition encompassing initiatives at the sub-regional level is an adequate starting point. The important point about local economic development is that it is usually locally led and delivered, even where it is using national resources.

Local economic development has its roots in the fundamental restructuring of both the national and international economies which took place between 1978–85. The major consequence of this was the rapid decline of the UK manufacturing sector, a result of worldwide over-capacity and a lack of competitiveness by UK firms. The decline had an effect throughout the UK, although urban areas such as the West Midlands and Newcastle upon Tyne were particularly badly affected. At the same time there was a corresponding rise in the service sector, although new jobs created did not match those lost in the manufacturing sector. Unfortunately, the skills requirement of the new growth sectors, and the locational requirements of the companies, has meant that inner city and urban locations were not always first choice. Urban centres no longer appear attractive to advanced manufacturing, routine office functions or warehousing and distribution.

At the same time, the UK elected a government which was, on the whole, non-interventionist. It believed that unemployment would eventually fall as a result of a strong and healthy economy, rather than through special measures. A belief that market forces would eventually lead to a strong economic recovery resulted in a reduction in the resources available for regional policy.

Although government-backed organizations such as the Manpower Services Commission and the Scottish Development Agency expanded their roles, there was a widely held belief that local action was required, and that additional resources had to be found. The leading bodies in local economic development were the local authorities, partly as a development of their industrial development role and partly as a response to central government's inaction.

Although local authorities were to lead in developing local economic strategies, they often followed, and were heavily influenced by, community based and community led initiatives. For a time, many authorities operated on an *ad hoc* basis. The idea of operating within an overall strategy came later. With the development of coherent strategies, the difference between local authority responses began to emerge. These have been categorized as non-interventionist, mainstream and radical interventionist.

Taking the 'non-interventionist' model of local economic development, it is true to say that development in the period between World War II and the 1970s occurred in the context of something of a bipartisan consensus concerning the objectives and techniques of macro-economic policy. This also extended to attempts to remedy geographical imbalances in the economy by means of centrally administered regional policies.

Generalizing across this period, the direct involvement of local authorities in the processes of local economic development was limited to planning and location-oriented activities in a facilitating, non-interventionist relationship with the private sector. In response to problems of declining traditional manufacturing industries in the late 1950s and early 1960s, additional powers to clear derelict land, provide industrial sites and give financial assistance to firms were granted to local authorities. However, it was not until the 1970s that involvement by local authorities in the local economy became widespread and widely based. Even so, such activity often remained marginalized and small scale due, among other things, to the traditional planning and property orientations of local authorities and the ensuring strength of the ethos that the impact of local authority actions upon the private sector should be minimized rather than maximized (Robinson, 1979).

Part Three examines public sector response at the local level to property development pressures and the opportunities, constraints and methods by which urban renewal of the physical fabric can be initiated and encouraged. To place this part in context and to give a wider perspective, this chapter provides an overview of local authority involvement in local economic and employment development of which property development is one aspect.

Indeed, a survey of local authority economic development activities for the Economic and Social Research Council (ESRC) (Mills and Young, 1986) showed that a model of, at best, passive facilitation of local economic development may have remained valid for many localities in the 1980s as far as local authority officers are concerned. If one includes within the model approaches to local development by councils such as 'having no strategy', 'relying on or co-ordinating the activities of other agencies', 'restraining or managing growth' or 'being generally supportive of economic development' without volunteering any specific activities in its favour, a considerable number of local authority respondents to the 1984 survey are found to conform to an arm's-length, non-activist model of local economic development (Mills and Young, 1986). Underlying such analysis is, however, the emergence in recent years of new non-local authority actors playing new local development roles.

Such a model in the 1950s and 1960s saw local economic development largely as an outcome of market forces and centrally determined and delivered government policy in which the local authority played a planning and facilitating role, provided social and perhaps some industrial infrastructure and extolled the locational advantages of the area. In the 1980s the relevance of the model persists in many areas with the proviso that the roles of central government macro-economic and regional policy have been downgraded and new supply-side central government policies (e.g. the activities of the MSC) delivered locally have entered the process, and new voluntary sector and private sector actors have emerged even in such traditionally 'non-activist' locations.

8.2 'Mainstream' model

More widely based, activist approaches by local authorities to local development emerged in the 1970s in response to widespread high levels of unemployment and progressive disillusionment with the ability of conventional macro-economic and regional policy to provide answers to new problems. The reorganization of local government in 1974 (1973 in Scotland) provided a stimulus to these developments with the creation of larger authorities, particularly in the metropolitan areas, with staff and resources for policy formulation. The local economy was identified as a key area for action by many authorities (Mawson and Miller, 1986). The principal areas of local economic development were the provision of sites and premises, financial assistance to industry, promotion and business development. The 1984 survey of Mills and Young found widespread expansion of the 'traditional' local authority role of sites and premises provision. Of all respondents, 82% had undertaken provision and/or improvement of sites and/or premises between 1980 and 1984. With the exception of a few very active authorities, this appears to have consisted of small scale activities concentrated at the lower end of the industrial and commercial property market with the provision of industrial and commercial units and serviced sites the most widespread (Chapter 11).

The second feature of the mainstream model of local authority economic activity, the provision of financial assistance takes the form of grants, loans, loan guarantees or in some cases equity. While initially controversial, this activity is widespread on a generally small scale. Similarly promotion of the attractions of an area either for inward investment or tourism remains widespread (70% of respondents), though again on a modest scale. Numerous local authorities claimed in the 1984 survey to be involved in business development, the definition of which encompassed the provision of information on sites and premises, business advice and support for enterprise trusts (Mills and Young, 1986). The above strands are the dominant themes of our mainstream model of local authority activity which has developed rapidly in the 1980s.

This approach by local authorities to local development has been characterized as essentially one of underpinning private enterprise (Boddy, 1984). The planning and land orientation of authorities noted above is reflected in a property-led approach dominated by the provision of sites and premises and with financial assistance still mainly linked to land and premises, although this has been loosened in many areas. Local authorities are seen as market and business related, seeking to stimulate or attract in private enterprise by creating favourable conditions for profitable investment, often in effect taking on some of industry's costs of production. The characteristic focus of the mainstream model has been the generation of local economic development upon the questionable assumption that this will directly transfer benefits pertaining to local employment without the need for any specifically targeted policies. This model places considerable emphasis upon new or existing small firms, partly reflecting the small scale of resources devoted by local authorities and partly reflecting a seedcorn approach to economic development.

The mainstream model of local development is one of intervention by local authorities which is essentially marginal, concentrated in the areas identified above and implemented with limited budgets and resources. Such provision has spread rapidly into rural and more prosperous regions from the areas of predominantly urban distress from which it tended to originate and has also been accompanied by the rise of actions by non-local authority bodies. The former point was amply illustrated by the 1987 survey on Economic Development Initiatives and innovations published by the Association of District Councils. This found widespread provision by district councils of premises and business advice, increased involvement with Local Enterprise Agencies and Trusts, considerable joint venture activity and about one third of respondents giving direct financial assistance (Association of District Councils, 1987).

8.3 'Radical–interventionist' model

In the early 1980s, a more strategic approach to local economic development emerged in a handful of Labour controlled authorities which corresponds to a 'radical–interventionist' model of local economic development. These approaches developed as a response to accelerating unemployment and posed a political and ideological challenge to the non-interventionist stance of central government. Their antecedents can be found within Labour Party debate in the 1970s; the concept of socially useful production pioneered by the alternative corporate plans of Lucas and Vickers *inter alia*; social audits of the community effects of closures; the revitalization of the co-operative movement; and Labour's embrace of a collection of disadvantaged local community groups (Mawson and Miller, 1986).

Radical, Labour-controlled authorities developed a pro-active role in the local economy bringing to bear a greater volume of resources and a wider range of initiatives than had hitherto been utilized by mainstream approaches. A growth in research into the pressures facing the local economy yielded policies targeted at specific locations, disadvantaged groups and sectors. This represented a considerable reaction to the trickle-down approach of generalized infrastructure and seedcorn provision of the mainstream approach.

The West Midlands County Council, the Greater London Council and the city councils in Sheffield and Leeds were pioneering authorities in this respect, all of them attempting to bring more resources to bear on the process of local economic development and to apply them in a more strategic, less defensive manner than the mainstream approaches. A rejection of the competitive auction for inward investment was mirrored by a commitment to endogenous economic development utilizing skills, resources and manpower within the local area invariably in sectorally targeted approaches to development.

The radical authorities were able to bring more resources to bear upon the process via a number of routes. One was simply the greater volume of resources concentrated within the metropolitan authorities. Their abolition has diminished and fragmented the resource pools available and has placed a greater onus on mechanisms adopted for the leverage of resources from outside of the radical authorities. One such method adopted was via Enterprise Boards which were created by a number of the authorities to intervene in the local economy and invest in local industry. Their roles ranged from one of fairly straightforward provision of finance to plug the investment gap, to investment in pursuit of wider social and economic objectives. As well as overcoming the constraints on local authorities holding majority stakes in companies, the Boards were able to mobilize private sector capital, and operate outside local authority bureaucracy and traditional orientations. Some of these implicit views on the shortcomings of local authorities *vis-à-vis* a strategic role in local development appear to have been shared by central government in a number of initiatives such as the establishment of Urban Development Corporations. The abolition of the metropolitan county tier has led to a reorientation of the activities of the Boards established by metropolitan authorities; the Boards are now seeking to find new resources and probably new roles as well, including co-operation with banks and venture capital funds.

Other authorities, notably Sheffield, did not operate via an Enterprise Board but sought to engage the private sector directly in partnership ventures. One of the reasons for this was that fewer resources were available at city rather than county level. The early attempts of Sheffield to engage the private sector were not particularly successful (Mawson and Miller, 1986) and led to the reorientation of policy towards bringing the authority's own role as an employer, purchaser and investor into line with the authority's objectives for local economic development.

A key feature of radical interventionist approaches is the pursuit of radical social objectives within the local development programme. While individual policies were and are targeted to perceptions of the local situation, they often include, *inter alia*, socially oriented conditions upon investment and planning agreements, support for co-operatives and community businesses, training initiatives in conjunction with the MSC, local colleges etc., and links with trade unions and the unemployed as an integral part of endogenous development policies. Technology transfer via collaboration with colleges and universities and the establishment of networks is also an important element of the endogenous approaches to development, as has been the use of purchasing power in contract compliance.

8.4 Local economic development initiatives

To examine the type of measures being undertaken within local economic development strategies, we can categorize initiatives under three broad headings: physical development, business development and people development. In practice some initiatives are aimed at more than one category.

Physical development

Within the physical development category are two distinct groups. The first is environmental improvement. In urban areas considerable resources have been devoted to environmental improvement in part to enhance the physical appearance of a particular area and also because such projects have been used to offer training/temporary jobs to unemployed people. A number of authorities have assisted individual projects to 'spin-off' community-based ventures from such schemes.

The second group of projects relates to the provision of workspace for small firms. This subject is dealt with in more detail in other chapters although it is worth repeating that the availability of large redundant buildings combined with a growing awareness of a significant gap in the property market led a considerable number of authorities to devote significant resources to workspace developments. Although the majority of developments were targeted at new and very small firms, a number of authorities focused their initiatives on particular groups, such as ethnic minorities.

Business development

Business development also breaks down into two broad categories, financial support and advisory services. Financial support is normally provided via loans and grants, although equity investment is also used. Nottingham County Council is a typical example of a mainstream authority. Loans are made from £5000 up to £0.5 million, repayable over 10 years; equity participation is available within the same financial limits for higher risk ventures; and grants of up to a few hundred pounds are available in special circumstances. In an evaluation exercise in 1986, some 30% of businesses receiving loans had failed, and 50% of those receiving combined loan/equity had failed. These figures are much higher than the banks would accept and also higher than venture capitalists would accept. This highlights the fact that local authorities involved in loan/equity support end up dealing with clients that conventional sources have rejected. Although some authorities set out deliberately to fill this gap, the consequence is a higher than average failure rate. The trend in financial support schemes is to part fund an activity rather than to offer a 100% subsidy. In particular schemes to encourage development (such as training) which will have a lasting effect on the competitiveness of the individual firm are becoming increasingly popular.

The second category within business development is advisory services. The provision of free advice, particularly for new starts, is seen as an important way of reducing the business failure rates. Increasingly business advice is being delivered through the 300 local enterprise agencies, autonomous organizations backed by both the public and private sector. Enterprise agencies were originally regarded with suspicion by some local authorities. Attitudes have changed, however, and the concept of using agencies to attract additional private sector support has gained widespread acceptance.

In the area of business advisory services, a number of initiatives aimed at special groups have been supported. A number of Scottish authorities have supported special advice units aimed at community business (and provided additional financial support). On similar lines, a number of Co-operative Development Agencies have been established. In most cases, the special support measures have been in addition to support for orthodox business advice.

People development

The final broad category of measures is people development. As unemployment has peaked and fallen, increasing attention has been focused on particular groups within the local labour market: women, ethnic minorities and the long-term unemployed. Demographic changes are leading to a reduction of the youth unemployment problem. These disadvantaged groups are often concentrated in inner city areas (or peripheral housing estates) and mainstream policies (of bodies such as the MSC) are perceived as inadequate. As a result, a number of local economic development strategies by local authorities, central government and other bodies target these groups.

The majority of measures relate to skills enhancement, either by offering places on specially designed training courses, or by offering employers incentives to hire and train people from priority groups. Many of these schemes have been supported by the European Social Fund. The whole concept of people development has moved forward considerably in the past five years. Less emphasis is placed on temporary employment.

One innovatory measure which has been closed by the Government is local labour clauses. A number of local authorities were concerned that local residents were rarely employed on public sector projects in their area. Policies which required successful tenderers to employ a certain percentage of local labour were initiated in London and Birmingham. This practice became illegal under the Local Government Act 1988, partly because the Government believed it clashed with a European Commission directive. Although local authority legal opinion disputed this, the practice of contract compliance is currently illegal.

One important development which is proceeding is the concept of targeted training linked to recruitment. The idea is that when a new development is announced, a local organization offers to train local people in the particular skills the development will ultimately require (e.g. retail). The employer is not obliged to hire people, although an interview is normally guaranteed. The idea is that the chances of local residents gaining employment is enhanced by undertaking a course relevant to the employment opportunities and the employer is offered a pool of labour trained in skills relevant to the particular industry.

Legislative changes

Two major legislative changes which will have a major effect on local economic development are currently being debated. An important legislative change is proposed in the Government White Paper 'The Conduct of Local Authority Business: The Government response to the Report of the Widdicombe Committee'. It is proposed to provide a general but limited power for local authority economic development activities and to limit the finance available for such activities.

The Widdicombe Committee considered the need to clarify the limits and conditions covering discretionary spending, including the use of sections 137 and 142 of the Local Government Act 1972. In 1984/85 two thirds of all discretionary spending was on economic development activities. Total expenditure by English local authorities in 1985/86 is estimated at between £280–£400 million, around one per cent of all local authority spending.

The White Paper recognizes the contribution which local authority economic development has made but states:

'There is no conclusive evidence concerning the cost effectiveness of local authority spending in this area, although in some areas local authority schemes appear to be comparable in terms of cost effectiveness to similar schemes funded by central government. It is difficult to judge whether local authority assistance results in investment that would not otherwise have taken place, or whether subsidies serve only to move investment from one area to another'.

Local authorities have never had a general power to engage in economic development. A number of specific powers relate to works on land advisory services. The use of section 137 has been an important element in 'plugging the gaps', to develop a coherent strategy. The Government intends to introduce a new specific power for economic development, and to outlaw the use of section 137 for economic development. It is anticipated, however, that the new legislation will reduce the range of activities which local authorities can undertake, and limit the amount of finance they are able to commit to economic development.

At the same time, the Department of the Environment has issued a consultation paper on local authority interests in companies (Chapter 12). The DoE put forward three categories of local authority interest or involvement in companies: local authority controlled, local authority influenced and minority interests. It is proposed to impose stricter controls on the first two categories and to specify in which types of companies a local authority can hold a minority interest.

The effects on enterprise boards and economic development companies may be considerable. If they remain local authority controlled, then it is likely that they will not be able to continue to make equity investment, and all capital expenditure will count against the parent authority's allocation. The result may be that local authorities will have to become a minority interest or to have the board defined as an 'arms length company'. The result of either action will be to reduce local authority control.

The response to the Widdicombe Committee and the DoE's Consultation Paper present major challenges to local economic development. The proposals will affect a number of authorities, although it is likely that strategies will be adjusted to minimize the effects of any new legislation, the role of local government in economic development could significantly be reduced. These changes reflect the Government's continued unease at the work being carried out at the local level, in spite of the lack of any evidence to prove that it significantly distorts competition or that the expenditure is not cost effective.

8.5 Trends in local economic and employment development

Over the past five years a number of important trends have emerged. The single most important trend is the concept of partnership between the public and private sector. Originally viewed with suspicion, the private sector is now regarded as a key actor at the local level. In part this represents public sector self-interest, as the private sector represents a new source of funds. It also, however, represents a recognition that the private sector has skills, knowledge, contacts etc. relevant to projects/initiatives. Partnership arrangements operate at two levels. The first is where companies are involved for philanthropic/social responsibility reasons; the second where they become involved for strictly commercial reasons.

Another noticeable trend is the increasing professionalization of staff involved in local economic development. There is now a large pool of professional individuals who have been involved in local economic develop-ment for a considerable period. This contrasts with the situation at the beginning of the 1980s where staff were recruited from diverse backgrounds such as estates management, planning and community work. The conse-quence of this change is that information exchange and informal networks for the transfer of experience are being developed. The overall result is to reduce the likelihood that the same mistakes are being made by similar type projects in different locations. A further result is likely to be that the local economic development lobby is likely to become more organized and more influential.

Another trend is the increasing importance of targeting. As local policies are being targeted at the small, disadvantaged groups within the local labour market, a major problem arises with the long-term unemployed in the 20–29 year old age group, who five years ago constituted the youth unemployment problem. In addition to the targeting of disadvantaged groups, there are a number of developments which seek to target disadvantaged areas. The most obvious example is the inner cities task forces. Perhaps a more typical case however is the peripheral estates of Glasgow, where a number of new initiatives are being developed. The concept of targeting relatively small areas, suffering from multiple deprivation, is likely to be a feature of local economic development in the 1990s.

It is worth noting at this point the proposed merger of the Scottish Development Agency and the Department of Employment's training function (the Training Agency). This will result in an all embracing economic development agency with an annual budget of £500 million. The suggestion is that the new body would operate through a network of about twenty local units (super-enterprise agencies). Each local 'office' would have considerable autonomy and would be governed by a board of directors, the majority of whom would be drawn from the private sector. The idea is to give economic development over to those who know best, i.e. businessmen.

Although there has been considerable criticism of the proposals, the idea of merging training with economic development has met with general approval. So also has the idea of locally based, autonomous organizations to 'deliver' policies. What has raised doubts is the lack of clarity regarding the role of local authorities and doubts about the private sector's willingness to become involved on the local network. The proposal seeks to introduce the American concept of private industry councils, run by 'prominent and concerned' local businessmen, mainly out of a sense of civic responsibility. Whether this idea can be transferred successfully to the UK is a matter of debate.

The interest in this development is that it indicates the current government thinking. The idea of combining all regional/national economic development activities, and then devolving power to local, business led organizations looks likely to be, eventually, applied to England and Wales, as already indicated in the Government's White Paper, 'Employment in the 1990s'. If the organizations are provided with adequate resources, and if local authorities and other community interests are fully involved, then this might present an effective way of channelling national resources into local economic development.

A further point of interest is the influence of US experience on local economic development. In certain areas, such as community led urban renewal, targeted training of disadvantaged groups and the involvement of the private sector, the US is more advanced than the UK. It is clear that US experience has and will continue to have a major impact on policy in the UK, e.g. job compacts, which originated in cities such as Boston. Due to the work of the European Commission, such as the LEDA (Local Employment Development Action) programme and the ELISE (European Information Network on Local Employment Initiatives) information network, there is now more exchange of experience between the member states, although language barriers still present problems. It is likely that as these networks become more established, ideas will be exchanged within Europe and will influence UK policy on local economic development.

8.6 Conclusion

Local economic development is a relatively new concept, although it is rapidly maturing. It has been pioneered, financed and led by local authorities, although legislative changes and an increasing local involvement by the private sector and central government are likely to alter this. There now appears to be an acceptance by both local and central government that local economic development is important, but that it should not be wholly financed by the public sector. In practice the mainstream model outlined earlier has been accepted by all sides as has the idea of a partnership between the public and private sector in local economic development. Increasingly emphasis is being placed on sharing experience, ensuring value for money, and developing monitoring and evaluation techniques. These changes reflect, in part, the success of the central government in changing attitudes about how and why the public sector intervenes and also reflect the maturing of the whole area. Perhaps the result will be that local economic and employment development will become much more systematic than in its early years but also that in the future new initiatives and local leadership will be coming from outside local government.

References

Association of District Councils (1987) *Economic Development Initiatives and Innovations*, ADC, London.

Boddy, M. (1984) Local economic and employment strategies. In *Local Socialism*, M. Boddy and C. Fudge (eds), Macmillan, London.

Business in the Community (1987), *The Contribution of Enterprise Agencies* (Report prepared by Enterprise Dynamics Ltd), BIC, London.

Chandler, J.A. and Lawless, P.L. (1985) *Local Authorities and the Creation of Employment*, Gower, Aldershot.

Chapman, J. (1986) *Venture Capital in the UK and its Impact on the Small Business Sector*, NEDO, London.

Directory of Social Change (1987) *Company Charitable Giving: 1987 Statistics*, Radius Works, London.

EEC (1987) *Social Europe: Action Programme on Local Labour Market Development (LEDA Programme), Supplement 3/88*, Commission of the European Communities, Directorate General V, Brussels.

Hausner, V.A. (ed.) (1987) *Urban Economic Change; Five City Studies*, Clarendon Press, Oxford.

Mawson, J. and Miller, D. (1986) Interventionist approaches in local employment and economic development: the experience of labour local authorities. In *Critical Issues in Urban Economic Development*, vol. 1, V.A. Hausner (ed.), Clarendon Press, Oxford.

Mills, L. and Young, K. (1986) Local authorities and economic development: a preliminary analysis. In *Critical Issues in Urban Economic Development*, vol. 1, V.A. Hausner (ed.), Clarendon Press, Oxford.

National Economic Development Council (1986) *External Capital for Small Firms*, NEDO, London.

OECD (1987) *Local Job Creation in the United States*, ILE Notebooks, OECD.

OECD (1987) *Decentralised Delivery of Manpower Measures*. Report by Evaluation Panel No. 8, OECD.

Robinson, F. (1979) *Local Authority Economic Initiatives: A Review*, Centre for Environmental Studies, London.

Sellgren, J. (1987) *Local Economic Development and Local Initiatives in the Mid-1980's*, Local Government Studies, Nov/Dec.

Steiner, M. (1985) *Old Industrial Areas: A Theoretical Approach*, Urban Studies, 22.

Storey, D., Keasey, K., Watson, R. and Wynarczyk, P. (1987) *The Performance of Small Firms*, Croom Helm, London.

Acknowledgements

The help of John McCreadie, until recently with The Planning Exchange and John Bowdery, The Management College Henley and IDP in the preparation of this chapter is gratefully acknowledged.

9
Government incentives to industrial and business space development

9.1 Introduction

The property development process, as an opportunistic response to a combination of physical, economic, social, environmental and political circumstances, is a high risk area of business activity. Fundamentally however, initiating a development scheme depends on the developer, singly, or in partnership with other agencies, preparing a detailed financial appraisal of the scheme in question. That calculation must reasonably assess the likely rental income that will be generated by the development and its probable capital value on completion and compare that to the expected costs of construction including borrowing, professional fees and land acquisition, in order to determine whether an acceptable profit or return on the investment will be achieved.

Calculations of this nature are clearly fraught with difficulties but assuming that the developer is satisfied that capitalized rental income less total costs will produce sufficient return, the implementation of the scheme will still depend on: (1) receiving planning permission from the local planning authority or the Secretary of State on appeal, and (2) persuading the landowner to sell the land at a price reflecting the developer's assessment of its potential. Even then, implementation may be withheld because of, say, unexpected market conditions or the possibility of incorporating higher earning uses in the scheme.

These circumstances inevitably mean that development proposals often fail to materialize. In the case of industrial developments where the market has, until recently, been characterized by high levels of vacant factory space left over from the economic recession, such failures have been commonplace and especially so north of a line drawn from the River Severn to the Wash, the so-called north–south divide. Similarly, new hi-tech and business space development schemes housing tenants involved in meeting the demands of increasingly technology-oriented markets, have tended to concentrate to the south of the same line, factors that have contributed significantly to maintaining long-standing regional economic imbalances.

Regional economic disparities are certainly supported by statistical evidence. Conspicuously, between 1979 and 1987, 28% of manufacturing employment has been lost, largely from those areas, regionally and within the inner cities, dominated by the most vulnerable industries. Indeed, Taylor observes during the same period, 94% of job losses have occurred north of the Severn–Wash line, where the northern regions have incurred a net loss of approximately 800 000 jobs, contrasting with the southern regions net jobs gain of 360 000 (Taylor, 1988). The differential impact of such trends on local unemployment is consistently and sharply demonstrated by official figures.

However, it is important to recognize that acute variations in the economic fortunes of the regions should not be overly generalized, including as they do pockets of severe depression south of the north–south divide, notably in the inner cities where, for example, some inner London boroughs suffer male unemployment in excess of 20% and, conversely, areas of success and prosperity to the north. Similarly, although questions of regional and inner area economic imbalance have historically been and continue to be proved, the contributory forces at work are dynamic and in some cases have brought about rapid and dramatic changes in the economic profile of a particular area in a short period of time. The recent experience of the West Midlands, for example, is indicative.

Nevertheless, overall, there is little doubt that measures of prosperity, physical standards and dereliction, economic opportunities and unemployment all confirm continuing and deep-seated regional economic disparities (Armstrong, 1987), and, that that situation is likely to worsen. According to the predictions of the Department of Land Economy at Cambridge University (DLE, 1987) in the decade 1985–95, of the 900 000 new jobs anticipated, almost half will be located in the south-east.

Such economic discrimination is of course not new. Many of the issues identified in 1934, when parliament first approved the Special Areas Act, the forerunner of the present system of regional aid, still remain. This continuing and apparently worsening regional economic dilemma prompted Leon Brittan, former Secretary of State at the Department of Trade and Industry, when addressing the Conservative Party Conference in 1985, to refer to such divisions between and within regions, as representing, '. . . one of the gravest social and economic problems facing us as a nation'.

The causes of spatial divisions in economic prosperity are widely recognized and reflect long-term international and national structural economic changes. Thus, for example, the shift in manufacturing emphasis to low wage countries especially in south-east Asia and the widespread rationalization by international companies of their least efficient and least profitable elements, have invariably adversely affected industrial activity in those already depressed areas of the United Kingdom, an outcome in part fuelled by long-standing but dated working practices, perpetuated by poor management and restrictive trade unions, which have contributed in limiting our competitiveness.

Meanwhile, within the country, historic and contemporary determinants of economic prosperity continue to favour the southern regions. Patterns of trade with Europe and beyond which have long benefited the south-east, as part of the 'golden triangle', are likely to be further concentrated with the removal of EEC trade barriers in 1992. New industries, founded on research and development investment and expanding service industries, have focused in the booming western and M11 corridors, the so-called 'golden horn' (Hall, 1988), while deregulation of the London financial markets and the City's half-way position between New York and Tokyo, have prompted enormous commercial investment in the City, with the inevitable multiplier effects on the employment and housing markets in the south-east. These factors all promote continuing disparity in the economic expectations of the regions perhaps best reflected in house price variations. Indeed, Robson (1988) notes that '. . . the cumulative effect of rises into the high 20% in the south-east and East Anglia, versus the less than 10% in the north – which may initially have been a reflection of the different regional economic prospects – have now become not an indicator but a perpetuator of disparity'.

Unfortunately, Government itself has added to this situation by:

1. concentrating infrastructure investment in the prosperous regions in response to demand. The third London airport decision, for instance, went ahead despite the efforts of back-bench tories from the northern regions desperate to attract such investment and their party colleagues' representing southern constituencies adamant opposition. Similarly, the completion of the M25 has generated enormous potential while the Channel Tunnel project recently prompted the Chairman of the North of England Regional Consortium to comment that, 'it is vital that both British Rail and the Government listen to the needs of passengers and businesses in the North if the Channel Tunnel is not to become another investment which fuels the economic divide between the regions'.
2. accepting a minimalist role for regional policies, where regional incentives to business activity are now generally worth less than in the past and where EEC regional funding has been pursued with less vigour than by other member states, notably the Republic of Ireland.

In fact, the Government's response, in pursuit of the monetarist doctrine, has been to focus on the growth areas on the basis that the 'ripple effect' will eventually benefit the depressed regions. Indeed, Kenneth Baker, then Secretary of State at the Department of the Environment, noted in a letter to the 1986 SERPLAN Conference, that by '. . . fostering its own economic growth, the south-east can act as a force for national economic recovery'. Accordingly, minimizing Government intervention, on the basis that it is an obstacle to development and thus economic growth, that unnecessarily restricts market forces, has resulted in the removal of such hurdles wherever possible.

It is clear that, in the short to medium term, the property market will actively respond to demand in the most buoyant areas, recently manifest in the enormous number of retail proposals, private sector new town projects such as Tillingham Hall and Foxley Wood, and office and business parks in urban fringe locations, which benefit from good accessibility, market proximity and international connections. What is far less clear is to what degree, if any, the ripple effect will benefit the depressed areas.

The Government is keen to identify hopeful signs and point out that the recent upturn in the economy has had some spin-off effects in the less prosperous areas, where unemployment statistics have shown slight improvements. Furthermore, Chartered Surveyors, Chesterton Bigwood, reported in early 1988, that,

'. . . last spring (1987) the northern tip of the north–south divide probably ended in a line from Bedford, Northampton to Banbury. But since that time, we have found demand for land and property increasing rapidly in the area to the north, with investment in retail, commercial and residential (NB not industrial!) vastly increased. We can now say that the northern tip of the north–south divide has moved to include the area stretching up to the Birmingham conurbation'.

But, not surprisingly, the outlying regions remain sceptical that any waves of economic prosperity, emanating from an overly-congested south-east, will reach them. Indeed, Robson (1988) suggests that,

'. . . rather than welcome a slow spread of the wave of growth, (the regions) more likely response will be to . . . fear the onset of a form of on-shore Taiwan in the north where, set against the high wage, high skill economy of the south, will be a low wage, low skill routine-production economy in the north, prone to reflect the ups and downs of economic fortune'

the 'two nations' view of the economy, while the poorer inner areas, including those in the south-east, view the prospects of continuing decentralization of people and jobs as a further threat to their economic revival.

In contrast to the advocates of the ripple effect, many, including the Institute for Employment Research (IER, 1987), believe that regional economic divisions are getting wider, that the new growth factors in the

British economy are working in favour of the southern regions and, conversely, the main factors of economic decay are affecting the outer regions of the United Kingdom. Indeed, most forecasts now accept that mass unemployment is here to stay, at least for the foreseeable future and that it will continue to discriminate on a regional basis.

Such views, however, are hardly likely to distract the Government from its economic strategy. But, having said that, although the Government may believe that market forces will eventually improve prosperity in the regions and the inner areas and therefore, there are no overriding economic reasons for actively pursuing an interventionist line, (though not all members of their own party accept that – see Brittan, 1988), there are good social and in particular political reasons for providing financial incentives to industrial and business space developments. Such inducements do of course amount to public subsidies and do run contrary to the Government's ideological and economic stance and thus only apply in selected areas or to particular schemes where they are implemented on a limited and tightly controlled scale. Nevertheless, incentives to promote developments do exist and this requires explanation.

Ironically, the main explanation stems from the fact that the southern regions and especially the south-east are themselves being disadvantaged as a result of the 'market knows best'. A hyperactive property market has not facilitated the establishment of comparatively low return industrial schemes, not least because of high development costs and a constrained labour market in which house price inflation is contributing to a serious skills shortage. Put simply, the prosperous regions are increasingly overcrowded and, together with the continuing decentralization of people and thus consumer spending power, the preference of entrepreneurs to locate in edge and out-of-town locations, thus benefiting from good accessibility and a pleasant environment, has inevitably increased the pressures on 'green' land. This trend has presented the Conservative Party with a potentially disastrous 'catch 22', in which free market monetarism, epitomized by the demand-led reaction of the development industry and especially the volume housebuilders, is in direct conflict with the ultra-protective stance of the Party's own grass root supporters anxious to protect their vested interests. This dichotomy has been highlighted by the Government's 'tightrope act' over Green Belt land releases and in its approval (or not) of restrictive Structure Plan reviews in the home counties. The recent controversy in Berkshire can only be repeated if the prediction of the Cambridge Econometrics Group (1987) comes to fruition, namely, that by the year 2000, more than one million people will leave the northern regions in search of jobs in the south, thereby fuelling still further pressures on the southern regions. Such circumstances lend credence to Brittan's view (1988), that, '. . . a policy that seeks to redress the balance between different parts of the country, is just as much in the interest of the overcrowded south-east of England, as it is of any other part of the country'.

Financial inducements to industrial developments must also be seen in terms of the increasingly sharp economic, social and thus political differences between and within the regions, divisions that were clearly demonstrated by the major political parties' constituency strongholds, following the 1987 General Election. In simple economic terms, the Government may view regional disparities and in particular, high levels of local unemployment, as acceptable casualties in contemporary Britain, but politically, and especially in the light of sporadic violent outbursts, notably in the inner cities and the public attention focused on these issues not least by prominent individuals and groups including Royalty and the Church, the Government must be seen to be addressing the problems, a challenge taken up by the Prime Minister herself following the election of her party for a third term.

It is in this context therefore, that it is important to appreciate the financial inducements that may be available to assist the implementation of industrial and business space developments, inducements that may persuade investors, developers and industrialists to pursue projects in areas and on sites that otherwise they would not have chosen.

9.2 Policy background

Apart from an early experiment aimed at promoting worker migration under the Industrial Transfer Act, 1928, all governments from 1934 to the present day have remained committed, albeit to widely varying degrees, to some concept of providing incentives to industrialists and developers, in order to attempt to counter regional economic imbalances. Indeed, policies founded on the notion of 'work to the workers', initially contained in the Special Areas Act, 1934 and then extended by the Barlow report's recommended pursuit of planned decentralization and dispersal of industrial population and economic activity in 1940, provided an underlying rationale, from which emerged a range of post-war economic, social and planning initiatives.

Those initiatives, invariably classified as 'stick and carrot' policies, were usually administered by central government and consisted principally of (a) restrictions on the further expansion of already prosperous areas, by, for example, limiting the issue of Industrial Development Certificates and Office Development Permits in congested areas, and (b) financial and other assistance in selected development areas to attract investment.

Immediately after World War II, such policies were reasonably successful in diverting industrial investment to the assisted areas, an improvement which prompted a relaxation of controls during the 1950s. But, by 1964 and in the light of national economic problems, areas dominated by, for example, heavy engineering and textiles, were again suffering from escalating unemployment, a trend which instigated a further period of active regional policy initiatives. The then Labour government established the

Department of Economic Affairs and formulated a National Plan, both of which were to prove short-lived and, in addition, created first, regional economic planning councils to prepare and implement regional strategies and secondly, a three-tier system of development areas, in which financial assistance was allocated on the basis of need.

Despite these efforts, however, and continued commitment to regional incentives until 1976 by both Labour and Conservative administrations, supplemented from 1975 onwards by financial inputs from the European Economic Community's regional development budget, unemployment rates, with marked regional disparities, continued to increase, trends which together with newly emerging factors were to prompt a major reassessment of regional policy. Contributory to that review were:

(a) adverse economic conditions nationally, in which rapid inflation and the intervention of the International Monetary Fund, necessitated reductions in public expenditure, including that allocated to regional policy vehicles, which, having peaked in 1975, have since been subject to dramatic cuts in available resources;

(b) increasing unemployment in the prosperous areas in the mid-1970s and especially in the inner areas of the conurbations (Dennis, 1978), where regional policy and the new towns programme, which encouraged the dispersal of population and employment over and above natural decentralis-ation, provided a plausible economic explanation and a convenient political scapegoat, a reaction which subsequently brought about a reallocation of resources. Indeed, the increasing emphasis placed on the economic fortunes of the inner cities, rather than the depressed regions, has generated a variety of financial incentives aimed at revitalizing the industrial and commercial characteristics of the designated areas. This approach contrasted with an initial concentration on the social and, in particular, racial issues in the inner cities, which stemmed primarily from concern about immigration in the late 1960s. Enoch Powell's infamous 'rivers of blood' speech in 1969 contributed in part to the launch of the first Urban Aid programme, enacted under the Local Government (Social Need) Act, 1969, which concentrated on special help for schools containing large numbers of immigrant children in Educational Priority Areas. This initiative, together with the Community Development Project, also established in 1969 and focusing on methods of supplementing social services in particular neighbourhoods, formed the basis of inner city policy but was soon overshadowed by issues of economic decline and physical dereliction. In 1972, Peter Walker, then Secretary of State at the Department of the Environment, initiated the Inner Area Studies to bring a 'total approach' to urban problems, based on a detailed analysis of six urban areas. Three were complete industrial towns – Oldham, Rother-ham and Sunderland – and three were districts of major cities, known to suffer from multiple deprivation, in Birmingham, Lambeth and Liverpool, where economic decline, physical decay and social problems were acute. The product of these lengthy and expensive studies, published between 1974 and 1977, formed the basis of the then Labour administration's white paper entitled 'Policy for the inner cities' (Oct. 1977), which became the Inner Urban Areas Act, 1978. The 1978 Act provided inner city local authorities, in partnership with central government, with extensive powers to pursue economic growth and in particular job creation, by direct and indirect assistance to investors and developers, with specific emphasis on the encouragement of small firms.

Reassessment of regional and inner city policy, was however, to take place in a very different political climate, following the Conservative's resumption of office in 1979 and from a revised economic stance, in which the Government, pledged to monetarism and public expenditure reductions to counter rising inflation, reacted to an increasingly severe economic recession with further public spending restrictions and a greater dependence on market forces to resolve economic and social issues. The Government's contribution to freeing up the market was epitomized in the title of the 1985 white paper concerned with reforms to the planning system, namely, 'Lifting the burden', which reflected the monetarist approach and involved removing yet more obstacles that constrained market forces including intervention by the Government itself and, in particular, by local government.

From that standpoint, several broad themes have emerged since 1979, which have fundamentally changed the nature and provision of financial incentives to industrial and business space developments. They can be identified as follows:

(a) the redefinition of areas designated for financial assistance. From broadbrush development area status which, at its peak in the early 1970s and, ironically, under a Conservative administration, involved half the nation's population living in regions eligible for aid, assistance has been more sharply targeted on much smaller geographic areas and, increasingly, on specific development sites. The Government has presented this approach as a more efficient method of focusing resources on particular growth points and thus generating economic activity from which further natural regeneration will multiply, while critics argue that this strategy is simply a means of minimizing the impact of substantial reductions in public expenditure. Whatever the motive, the outcome has been manifest in two ways:

1. development areas which qualify for grant aid have been reduced in scale and status. Although subject to review in August, 1982, the three-tier development area system – that is special development, development and intermediate areas – was retained until November, 1984, when substantial changes, initiated by Norman Tebbit during his brief reign as Secretary of State at the Department of Trade and Industry, were implemented by Norman Lamont. Special development areas were

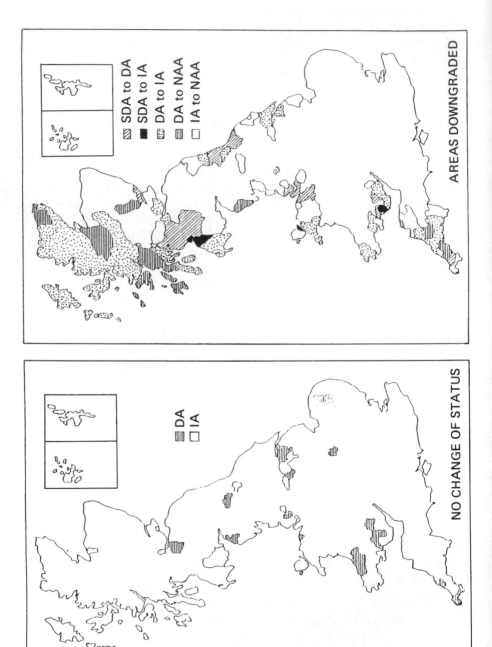

AREAS DOWNGRADED

SDA to DA
SDA to IA
DA to IA
DA to NAA
IA to NAA

NO CHANGE OF STATUS

DA
IA

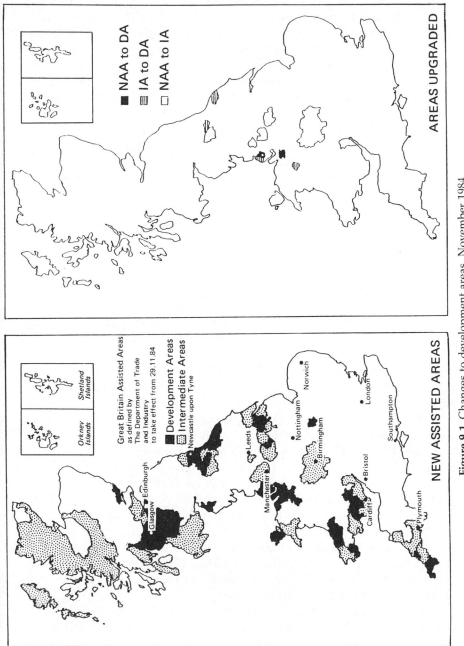

Figure 9.1 Changes to development areas, November 1984.
Source: Planning 600, 11 January 1985.

abolished and the new development and intermediate areas, in which the former qualified for automatic Regional Development Grants and discretionary Regional Selective Assistance and the latter for RSA only (both detailed later), were redefined on the basis of first, 'travel-to-work-areas' in which unemployment blackspots, located close to or within more prosperous areas, were excluded from designation and secondly, by the Government's indexation of need, formulated by reference to selected employment criteria. The result, illustrated in Figure 9.1 shows some extension of intermediate area status to include, for example, the West Midlands, which thus qualifies for European Regional Aid but, overall, a reduction in the areas able to receive Regional Development Grant, from a previous coverage of 22% of the working population, to 15%, involving 87 areas being downgraded or removed from assisted area status.

2. in addition to extending the list of designated districts under the Inner Urban Areas Act, 1978, which may then benefit from the Urban Programme, new designations have been introduced which focus fiscal and/or administrative concessions on tightly defined areas with particular economic problems. Thus, the Local Government, Planning and Land Act, 1980, provided the Secretary of State at the Department of the Environment with the power to establish Enterprise Zones, since widely utilized, while in 1984 Free Port Zones became official policy, with the establishment of six experimental areas.

(b) increasing centralization of administrative and budgetary control. Since 1979, the role of local authorities as the vehicle through which particular projects have been identified and subsequently financed, has been reduced, while the Regional Economic Planning Councils, initiated by Lord George Brown in 1964 and responsible for producing well researched regional strategies, had no meaningful powers of implementation or funds and were abolished by the Conservatives shortly after coming to power in 1979, reflecting the government's general aversion, at that time, to the extensive range of quasi-autonomous, non-governmental organizations (quangos). Ironically, although particular ministries, notably the Department of Trade and Industry and the Department of the Environment, have extended their direct project and financial control powers, the Government have also established new quangos, in particular the Urban Development Corporations, which are responsible directly to central government and by-pass the traditional role of elected local authorities.

(c) the introduction of the concept of 'leverage'. The impact of public expenditure cutbacks, especially since 1979, on regional aid and to a lesser extent on the inner city programme, where apparent increases in funding have been significantly undermined by adverse rate support grant settlements and rate-capping, has inevitably necessitated greater reliance being

placed on attracting private sector investment in order to secure implementation of much needed, but financially unattractive development projects.
The Government has therefore adopted, wherever possible, the 'leverage'
strategy, that is, allocating to particular schemes, the minimum amount of
public funding necessary to secure the maximum input of private sector
finance, an approach well illustrated in the use of Urban Development
Grants and Urban Regeneration Grants and their replacement, the new City
Grant.

In the light of these basic themes, it is now appropriate to detail the range
of government grants and incentives that may be available to developers to
facilitate the implementation of industrial and business space projects.

9.3 Government grants and incentives, 1987/88

A synopsis of government grants and incentives, available in the financial
year 1987/88, is tabulated in Figure 9.2 for ease of reference and comparison,
and explanatory notes are provided below. It should however be emphasized that Derelict Land Grant, Urban Development Grant and Urban
Regeneration Grant were superseded by a unified City Grant in the financial
year 1988/89, while Regional Development Grant aid has also been subject
to change. These modifications are detailed in section 9.4.

Derelict Land Grant

The Derelict Land Survey, 1982, revealed 46000 hectares of derelict land in
England, 34000 of which were viewed as justifying reclamation. Priorities
for reclamation were defined in DoE Circular 28/85 – Reclamation for Reuse
of Derelict Land – which stressed the importance of schemes designed to
restore land, especially in the inner
commercial uses. As a result, in addit
designated under powers contained in
were further focused on selected are
three-year rolling programmes were
constraints imposed by funding on an

Derelict Land Grant aims to assist
ranges from the remnants of mine
industrial processes and, since 1979,
hectares of land. To be eligible, land
and be incapable of beneficial use w
regionally and are available to loca
including the nationalized industries
and individuals.

Figure 9.2 1987/88 Go
Source: Pepper, Angliss and

	Derelict Land Grants	Urban Development Grants	Urban Regeneration Grants	Urban Development Corporations
1. Sources of Aid.	Dept of the Environment (via Local Authority) Room P2/109.	Dept of the Environment (via Local Authority) Room P2/127.	Dept of the Environment Room P2/126.	London Docklands Merseyside Trafford Park Black Country Teesside Tyne and Wear
2. Objectives and scope	The reclamation of derelict land to bring it back into beneficial use or to improve its appearance.	To promote the economic and physical regeneration of deprived urban areas by encouraging development to strengthen the economy and bring land and buildings back into use. Applies to commercial, industrial, warehouse, residential, leisure, tourist and hotel developments; both new build and conversion.	To promote the economic and physical regeneration of older urban areas, severely affected by industrial change. Applies only to large sites typically 20-100 acres and large groupings of derelict and run-down buildings of at least 250 000 sq.ft.	To regenerate the designated area through reclaiming and servicing land; renovating buildings. Providing basic infrastructure in order to encourage development of industry and commerce.
3. Amounts allocated from 1987/88	£81m.	£30m (shared between UDGs and URGs).	£30m (shared between URGs and UDGs)	£126m.
4. Types of aid	A grant based on a percentage of the net loss to the owner as a result of carrying out the work i.e. the cost of work less the increase in value due to the work.	Grants and/or loans in order to make the project viable, or to bridge the gap between cost and completed value.	Outright or repayable grants, or loans.	Discretionary aid by UDCs to provide assistance to developers in suitable cases.

vernment grants and incentives for developers in England.
' Yarwood (1987).

Enterprise Zones	Urban Programme (Inner Urban Areas Act 1978)	Regional Development Grants	Major Projects Section 8 of Industrial Development Act 1982	Land Registers
Dept of the Environment Room P2/108.	Designated Local Authorities	Regional offices of Dept of Trade and Industry.	Dept of Trade and Industry Room 336.	Dept of the Environment Room P2/108.
To stimulate economic activity within designated zones through lifting certain financial burdens and administrative controls.	To help tackle the economic environmental and social problems of inner cities.	To encourage approved projects to be undertaken in the Development Areas.	To enable major large projects of national interest to be undertaken which would, in absence of grant, not be possible.	To publicize the existence of unused land owned by the public sector. Register can be inspected, free of charge, and contains details of ownership, history and constraints.
N/A	£2994m.	£300m (shared with other DTI assistance schemes).	£300m (shared with 3 other DTI assistance schemes).	N/A
Exemption from rates; 100% tax allowance on capital expenditure on industrial and commercial development; a simplified planning regime; reduced administrative requirements.	Loans and grants for environmental improvement, land clearance and site preparation, landscaping, infrastructure works, conversion of buildings for commercial use.	Grants towards job creation through investment in new buildings, works and plant and machinery.	Grants towards cost of fixed capital in buildings and plant and machinery (and exceptionally other related items).	Minister has powers to direct the owner to dispose of the land or to require owner to record unlisted sites on the register.

(**Figure 9.2** continued)

	Derelict Land Grants	Urban Development Grants	Urban Regeneration Grants	Urban Development Corporations
5. Areas eligible	80% in Assisted and Derelict Land Clearance Areas: 50% in all other areas.	Priority is given in the following areas: Barnsley, Birmingham, Blackpool, Bolton, Bradford, Brent, Bristol, Burnley, Coventry, Derby, Doncaster, Dudley, Gateshead, Greenwich, Hackney, Halton, Hammersmith & Fulham, Haringey, Hartlepool, Islington, Kensington & Chelsea, Kingston-upon-Hull, Kirklees, Knowsley, Lambeth, Langbaurgh, Leeds, Leicester, Lewisham, Liverpool, Manchester, Middlesborough, Newcastle, Newham, North Tyneside, Nottingham, Oldham, Plymouth, Preston, Rochdale, Rotherham, St. Helens, Salford, Sandwell, Sefton, Sheffield, South Tyneside, Southwark, Stockton, Sunderland, Tower Hamlets, Walsall, Wandsworth, Wigan, Wirral, Wolverhampton, Wrekin	All areas eligible but priority is given to the most deprived areas with severe unemployment and substantial amounts of disused industrial and commercial property.	Only within designated UDCs (see above).
6. Limitations and notes	Costs relating to land purchase, infrastructure works and administrative expenses are included.	The developer must meet the greater part of the costs involved. Local Authority must support application and pay 25% of grant.	Not available for schemes which appear feasible without a grant or loan. Grants are paid direct to the developer by DOE.	These are entirely discretionary financial grants and other forms of assistance. (UDCs are also the planning authority for their designated area.)

Enterprise Zones	Urban Programme (Inner Urban Areas Act 1978)		Regional Development Grants	Major Projects Section 8 of Industrial Development Act 1982	Land Registers
Corby	Birmingham	Leicester	The Development Areas.	Anywhere in England.	Throughout England.
Dudley	Hackney	Lewisham			
Glanford	Islington	Middles-			
Hartlepool	Lambeth	brough			
Isle of Dogs	Liverpool	Newham			
Middlesborough	Manchester	North			
NE Lancashire	/Salford	Tyneside			
NW Kent	Newcastle/	Nottingham			
Rotherham	Gateshead	Oldham			
Salford/Trafford	Barnsley	Plymouth			
Scunthorpe	Blackburn	Preston			
Speke (Liverpool)	Bolton	Rochdale			
Telford	Bradford	Rotherham			
Tyneside	Brent	St Helens			
Wakefield	Bristol	Sandwell			
Wellingborough	Burnley	Sefton			
Workington	Coventry	Sheffield			
	Derby	South			
	Doncaster	Tyneside			
	Dudley	Southwark			
	Greenwich	Stockton-			
	Halton	on-Tees			
	Hammersmith	Sunderland			
	& Fulham	Wrekin			
	Haringey	Tower			
	Hartlepool	Hamlets			
	Kensington	Walsall			
	& Chelsea	Wandsworth			
	Kingston-	Wigan			
	upon-Hull	Wirral			
	Kirklees	Wolver-			
	Knowsley	hampton			
	Langbaurgh				
	Leeds				

Benefits are limited for 10 years from the date of designation of the EZ in which the project is located. A balancing tax charge is made if buildings are sold within 25 years.	Assistance is limited to 50% of the cost. Projects must improve the employment prospects and/or reduce number of derelict sites and vacant buildings within the designated area. First priority to Partnership Areas; second Designated Authorities; third other non-designated local authorities.	Land, second hand assets and vehicles not eligible for capital grants. Grant is restricted to higher of: 15% of eligible capital expenditure or £3,000 for each new job created. Further limitations apply where more than 200 new jobs are created; if under 200 new jobs limitations also apply where the capital expenditure is more than £500 000.	Minimum new investment to be at least £500 000 and projects must meet a range of stringent criteria laid down by DTI.	Sites under lease and which are to be used within 2 years are excluded. Also land used for public open space, allotments, recreation, agriculture. Defence land and Government property are not included.

Awards are determined on the basis of demand for sites, value for money and technical feasibility, grants being paid as a proportion of the cost of bringing the land back to its natural state. In assisted areas and derelict land clearance areas, local authorities receive 100% of the costs incurred, while private firms, voluntary organizations and other public bodies receive 80%. Outside such areas, the rate is 50% except in national parks and areas of outstanding natural beauty, where local authorities receive 75%.

During 1987/88, the Government made a total of £81m available, £71m to local authorities and almost £10m to the private sector scheme, an input which to date, on industrial and commercial sites has achieved a public/private sector gearing ratio of 1:10. For 1988/89 however, total funding will be £76m, returning to the 1985 level, while the number of areas with rolling programmes will be reduced from 16 to 13.

Derelict Land Grant has certainly succeeded in encouraging reclamation but this must be viewed in the light of increasing rates of dereliction during the 1980s, especially since the economic recession and a budget allocation which peaked in the mid-1980s and is now declining. Concern has also been expressed at the proportion of reclaimed sites that remain undeveloped and the rate of restoration which has failed to increase. In fact, the figures for 1985/6 and 1986/7 (estimated) at 1 060 and 1 114 hectares, were the lowest since 1979, reflecting in part the shift in emphasis from relatively simple projects, such as the upgrading of waste tips to public open space, to more complex and expensive reclamation of old industrial sites to a standard suitable for redevelopment. That concern was confirmed and expanded upon in a report by the National Audit Office, published in November, 1988, which, while recognizing the benefits of reclamation, seriously questioned the impact of grants on investment decisions and was especially critical of the DoE. The ministry was accused of inadequate appraisals of projects, insufficient checks on the reliability of applicants' information and no regular monitoring of intended and actual end uses. The clear implication was that Derelict Land Grant was open to abuse.

Most recently, as part of the policy review contained in *Action for cities* (HMSO, 1988), the private sector element of Derelict Land Grant is to be replaced by the new City Grant.

Urban Development Grant

The main objective of the Urban Development Grant (UDG) scheme, when introduced into the Urban Programme in 1982, was to stimulate the economic regeneration of designated districts and enterprise zones by 'levering' significant private sector investment into inner city projects that would not otherwise be viable for companies to pursue. Thus, government commit the minimum public sector contribution necessary to bring forward otherwise marginal private sector development schemes, where their implementation

will contribute to economic development by generating employment opportunities for inner city residents. In order to qualify for the grant, the
developer must show that the project meets social needs in the inner area
and that, in particular, the amount of grant sought is the minimum necessary to make the scheme profitable.

The UDG development process is initiated by the developer identifying a
potential scheme or being invited to produce proposals by the local
authority. An application is then submitted to the authority which acts as
sponsor and, jointly, an approach is made to the DoE, where a specialist
appraisal team seconded from the private sector, vet the proposal in terms
of its social benefits and financial implications, the latter normally being
considered in the form of a residual valuation. The developer must justify
his financial calculation and where appropriate, UDG will be offered,
although claw-back provisions may be attached to offset underestimated
returns or overestimated costs. Up to 75% of the grant aid is provided by
central government and the remainder by the local authority.

According to the DoE sponsored evaluation of the UDG programme
(HMSO, 1988), by June 1986, UDGs totalling £78m had been approved for
177 projects and a public/private sector investment gearing ratio of 1:4.5 had
been achieved. Projects are not restricted in type or scale and in practice,
have proved extremely varied though, of those evaluated in 1986, the
majority consisted of commercial and industrial developments, where
£34.9m and £25.6m of grant aid attracted £195.3m and £100.7m of private
sector investment respectively. New and refurbished industrial schemes
have been completed, mainly providing small workshop accommodation,
but new employment opportunities have been fewer than anticipated, with
costs per full-time job ranging from £12000 to £17000 overall (Pearce, 1988).

Urban Development Grant has certainly succeeded in attracting private
investment to the inner city, from which employment and environmental
benefits have emanated, but the level of activity in investment terms has
been relatively small. Total grant aid has consistently fallen short of Urban
Programme allocations, where high failure rates and low overall numbers of
applications continue to reflect the long-standing caution of the financial
institutions to such projects. Perhaps inevitably therefore, UDG successes
have tended to concentrate in areas where market conditions, local political
commitment and officer skills have cultivated development opportunities,
notably in Birmingham, Leeds, Nottingham and Dudley. However,
protracted appraisal periods, during which time market uncertainties may
emerge, prompted calls for review and the Government responded in its
booklet, *Action for cities* (HMSO, 1988), detailed in section 9.4.

Urban Regeneration Grant

Launched in April, 1987, under enabling powers contained in the Housing and Planning Act, 1986, the Urban Regeneration Grant is also designed to attract private investment to areas affected by the decline of traditional industries and thus given priority under the Urban Programme. The grant, unlike UDG, is awarded direct to the private developer, by-passing local authority involvement and focuses on substantial sites of more than 20 acres or buildings with more than 250 000 sq. ft of floor space, where potential residential, commercial, industrial or leisure development, which will strengthen the economy of the area, cannot be implemented without financial support. The initiative is therefore aimed at medium sized areas of derelict or neglected land and is intended to complement smaller UDG projects and the large scale activities of the urban development corporations.

URG is available in various forms. It may be paid directly as a grant, as a loan or be subject to conditional repayment, the terms of which are negotiable. By the end of 1987, two awards had been made, one involving a grant of £3.25m to assist in reclamation work and the provision of infrastructure on the 36 acre site of the former Round Oak Steelworks, located within the Dudley Enterprise Zone, where a £17m development will provide 421 000 sq.ft of industrial and commercial units, which could generate 1 000 jobs. However, although other schemes are under consideration, URG will also be superseded by the new City Grant.

Urban Development Corporations

Following its election to office in 1979, Mrs Thatcher's administration expressed its distaste for the 'concealed growth of government', that is, quasi-autonomous, non-governmental organizations or quangos and pledged to reduce their number. Sir Leo Platzky's White Paper, published in 1980, focused on such bodies and, by 1984, their number had been reduced from 2 167 in 1979, to 1 653. In contrast however, this apparent aversion has not applied to planning and development issues, especially in the inner cities. Indeed, the qualities embodied in these organizations, including their ability to focus single-mindedly on particular policy issues and provide an integrated approach to implementation with the aid of specialist enterprise, together with the Government's ability to control their activities and budgets directly, without local interference, has proved increasingly attractive.

The Local Government, Planning and Land Act, 1980, introduced the concept of the Urban Development Area and enabled the Environment Secretary to proceed with their designation in order to 'secure the regeneration of the area' (S. 136, 1980 Act). To facilitate the inner city revitalization process, the Secretary of State is empowered to establish an Urban Development Corporation and appoint a board to manage its operations and achieve its objectives by bringing land and buildings into effective use, encouraging the development of industry and commerce, creating an attractive environment and ensuring that housing and social facilities are available. To do so, the corporation has general powers to acquire, compulsorily if necessary, hold, manage, reclaim and dispose of land and carry out building and infrastructure works. Additionally, the Secretary of State can confer on a UDC the town planning and housing responsibilities, normally vested in the duly elected local authority, while building control and certain highway and public health functions can also be transferred.

The London Docklands and Merseyside Development Corporations, launched in 1981, were the only UDCs until 1987, when the Black Country, Trafford Park, Teeside, Tyne and Wear, and Cardiff Bay corporations were also established and three mini-UDCs were designated in Bristol, Leeds and Central Manchester. Finally, in March 1988, a development corporation was created in Sheffield and this, according to the Environment Secretary, represented the end of the current phase, since there would be insufficient funds for any further major additions in the foreseeable future. The new breed of mini-UDCs, each with a budget of up to £15m available over four to five years, have been identified because they all have significant amounts of derelict land concentrated in small areas, designations varying in size from 250 to 1 600 acres.

Urban development corporations are financed primarily by Exchequer grant-in-aid but additional funding can accrue in the form of capital receipts emanating from acquisitions and disposals of the corporation's property portfolio. Small amounts of borrowing may also be available from the National Loans Fund. Table 9.1 details current and planned expenditure on Urban Development Corporations (£ million):

Table 9.1 Current and planned expenditure on Urban Development Corporations

	1986–87 outturn (£m)	1987–88 outturn (est.) (£m)	1988–89 plans (£m)	1989–90 plans (£m)	1990–91 plans (£m)
Public expenditure provision	89	133	203	215	223
Estimated receipts net of corporation tax	23	43	121	88	63
Total resources	112	176	324	303	286

Clearly, significant public funds are now channelled through urban development corporations. Their approach to implementation, however, rests on creating the confidence and long-term investment climate which will attract the funding institutions and, therefore, much of their public money is invested in infrastructure and environmental improvements in order to facilitate the process of promotion and marketing. Nevertheless, grants and other forms of assistance are made available at the discretion of the corporation to the promoters of particular development projects, which may, for example, take the form of a partnership arrangement in which the corporation utilizes its land assembly powers and makes the land available on favourable terms to a developer, who, as a result, is able to proceed with an otherwise impractical scheme. Such arrangements are site specific and subject to individual negotiation.

Since their inception, urban development corporations have been consistently open to the charge that they have not resolved the employment and housing problems of the indigenous population, in part because the corporations are not governed by locally elected representatives and are not therefore seen to reflect local views and, further, as a result of the particular impact of the London Docklands Development Corporation. The enormous success of the LDDC in attracting new commercial investment and luxury housing developments has rarely benefited the local population, despite an investment of £400m of public money. Indeed, it has brought about spiralling land prices which have encouraged the out-movement of local businesses, on occasion via the LDDC's compulsory purchase powers, in order that escalating development values can be released. Although impacts in the other urban development areas are unlikely to be as dramatic, the LDDC's track record has inevitably polarized views on the efficacy and equity of this approach. Furthermore, recent accusations of financial mismanagement involving overspending and poorly administered land deals, prompted Michael Heseltine who formulated the LDDC, to call for a ministerial inquiry into its operation and the National Audit Office to recommend more stringent controls.

Enterprise zones

The question of unemployment black spots in inner city areas, characterized by structural decline in the local economy and environmental dereliction, has of course been subject to considerable government attention in planning and development terms. Enterprise zones, the 'flagship of deregulation' since 1979, were first suggested by Peter Hall in 1977 and subsequently featured in Sir Geoffrey Howe's 1980 budget. In March, 1980, an explanatory document, entitled Enterprise Zones Policy Proposals, was published and defined the purpose as, '. . . to test as an experiment and on a few sites, how far industrial and commercial activity can be encouraged by the removal of certain fiscal burdens and by the removal or streamlining of certain statutory or administrative controls'. In particular, the aim of setting up an Enterprise Zone is to achieve a significant impact in terms of new development, improvement of existing property or increased economic activity, within a ten year timescale.

To create conditions that will favour new investment, the following financial and administrative incentives apply within the designated areas:

1. Exemption from the payment of rates on industrial and commercial property, whether new or existing, for a ten year period. This applies only to the general rate.
2. 100% allowances for corporation and income tax purposes, for capital expenditure on industrial and commercial buildings, subject to a drawback provision if the premises are sold within 25 years of construction.
3. A simplified planning scheme would operate whereby developments complying with the published scheme for each zone would not require planning permission. In effect, the scheme is a broad-brush plan which indicates classes of development permissible within the zone and determines acceptable standards of health, safety and pollution. Controls remaining in force would be administered more speedily.
4. Government requests for statistical information would be reduced.

At inception, developments in enterprise zones were exempt from Development Land Tax and Industrial Development Certificates were not required. Both have since been abolished nationally.

There are 25 enterprise zones in the United Kingdom, 17 in England, 3 each in Scotland and Wales and 2 in Northern Ireland. The first 11 zones were established between June 1981 and March 1982 while a further 14 areas were identified between July 1983 and April 1984. The latest zone, embracing Greenock on the Lower Clyde where the Scott Lithgow shipyard is being mothballed and male unemployment is approaching 30%, brings the total to 26. However, in December 1987, Nicholas Ridley, Secretary of State at the DoE, announced that the Government did not intend to designate any further zones in England, although new and extended zones will be considered elsewhere in the UK, because other measures, including urban development and regeneration grants, were regarded as more cost effective ways of stimulating jobs and investment in deprived areas. This conclusion stemmed from the Government commissioned evaluation of enterprise zones and the annual monitoring statistics, which revealed that 35000 of the 63300 jobs created in the 23 British zones (Northern Ireland is considered separately) were a direct consequence of EZ policy. It was further calculated that 13000 net additional jobs were supported directly and indirectly in the local economies influenced by the zones, but the cost per job created was estimated at between £23000 and £30000, a level of expenditure which clearly influenced the Government's review of zones where public financial input overall now exceeds £300m.

In addition, considerable doubt has consistently surrounded the nature and impact of employment generated in the zones, many existing firms relocating in order to gain a financial advantage over their competitors but, as a result, only contributing to a realignment of unemployment patterns. Furthermore, in some industries, competing companies outside the zones have been seriously undermined, a situation which prompted a landmark decision in the case of Clement (VO) v. Addis Ltd (House of Lords, 11/2/88), where the House of Lords held that a factory in the Swansea valley should receive a rate reduction because the neighbouring enterprise zone amounted to a change in the state of the locality and thus affected rateable values. However, the likelihood of a flood of similar claims around EZs was quickly curtailed by the Environment Secretary amending the Local Government Finance Bill, to prevent rateable values being changed in advance of the general revaluation in 1990.

Nevertheless, although the employment repercussions of enterprise zones are questionable, there is little doubt that their designation has promoted considerable development, especially in the commercial sphere where vast schemes such as Canary Wharf in the Isle of Dogs EZ and the MetroCentre in the Gateshead EZ, qualify for financial inducements that would not be available for such projects outside the zones.

Urban Programme

As noted, the Local Government Grants (Social need) Act, 1969, introduced a small Urban Aid programme, which, following the enactment of the Inner Urban Areas Act, 1978, was renamed the Traditional Urban Programme. More importantly, the 1978 Act introduced the Urban Programme which has formed the backbone of inner city policy to the present day, having been retained by the Conservatives since their election success in 1979.

The Programme aims to assist local authorities in responding to the economic, environmental and social repercussions of long-term structural changes in the economies of older towns and cities, by fostering new enterprises and upgrading the environment in order to promote investment.

The Act empowers the Environment Secretary to identify particular inner city areas, which on the basis of selected criteria, suffer from multiple deprivation. These areas, known as designated districts may;

1. Declare improvement areas, either industrial or commercial in character, in which they can make loans or grants for environmental improvements and give grants for the conversion and improvement of industrial or commercial buildings
2. Make loans on commercial terms for the acquisition of land and for carrying out building or site works
3. Give loans or grants towards the cost of setting up common ownership or co-operative enterprises.

Designated districts, until the financial year 1986/87, were given priority in the order of the following categories:

(a) Partnership areas, designated on the basis of the severity of the problems faced, prepare an Inner Area Programme (IAP) annually on a three year rolling basis, which is agreed by a Partnership Committee, convened by the Department of the Environment and involving representatives of the public, private and voluntary sectors and processed through the Urban Programme Management Initiative, launched in 1985. The IAP, prepared by the local authority takes the form of a bid for resources for particular projects.

(b) Programme authorities are also areas where special aid is necessary and they too prepare an Inner Area Programme but subsequently follow a simplified administrative process.

(c) Other designated authorities and selected authorities without programme area status, were invited to bid for resources from the Traditional Urban Programme.

In addition, the 1978 Act provides partnership areas with extra powers to promote economic regeneration, namely rent support to companies taking new leases on industrial or commercial premises, interest-free loans for site preparation works and interest-relief grants to small firms for loans on land and buildings.

In 1987/88 however, this structure was simplified and all designated authorities must now submit a co-ordinated IAP, while the Traditional Urban Programme was phased out.

Initial allocations are subject to the formal assessment and approval of the IAP and the provision of a satisfactory report of the achievements of the programme in the previous year, for each designated district. In February 1988 therefore, the Government announced initial Urban Programme funding of £259m for the year 1988/89, to be divided between 57 partnership and programme authorities. Local authorities receive 75% grant aid from central government in support of approved projects, which often involve partnership with other agencies, including the private sector. Assistance to developers is limited to 50% of the costs involved on projects which must improve employment prospects and/or reduce the number of derelict sites and vacant buildings within the designated areas.

The Department of the Environment review (HMSO, Oct. 1987) of the implementation of the Urban Programme for the year 1985/86, reveals the type of projects funded, one third of which were in pursuit of economic objectives, in which the provision of land and buildings for industrial development were prominent.

The largest single project to be funded by the Urban Programme, the Greater Manchester Exhibition Centre was completed in 1986 and involved the conversion of the former Central Station into 10 350 sq. m of exhibition and conference space and 1 500 car parking spaces. Half the estimated conversion costs of £17.8m were provided by the Urban Programme, while

Table 9.2 Resources allocated to The Urban Programme since 1979

	1979/80 (£m)	80/81 (£m)	81/82 (£m)	82/83 (£m)	83/84 (£m)	84/85 (£m)	85/86 (£m)	86/87 (£m)	87/88* (£m)
Partnerships	110	116	118	143	133	131	125	122	120
Programme authorities	32	47	50	77	86	97	96	103	140
Other designated districts	3	6	5	9	8	8	10	9	—
Traditional Urban Programme	30	33	42	47	53	47	44	37	10
Other	10	—	—	19	30	38	42	41	54
Total	185	202	215	295	310	321	317	312	324

*Initial allocation.

higher final costs were in part supported by a grant of £4m from the European Regional Development Fund. Private sector investment exceeded £9m.

The track record of the Urban Programme is impressive with over 12 000 specific projects benefiting from grant aid but, on closer examination, the impact of the programme and, in particular, the real level of funding available, is less convincing. Critics have argued that a variety of fiscal and financial measures work in direct opposition to the interests of the Urban Programme and, that while one of the principal aims of inner city policy had been claimed to be the targeting of assistance to those most in need, there is little explicit policy which appears to be specifically directed to this end; for instance new jobs do not necessarily benefit local residents (ESRC, 1985).

Certainly, emphasis on the Urban Programme, utilizing local authorities and involving major public sector financial inputs, has waned in the face of other initiatives that are more dependent on private investment and, increasingly, subject to direct approval from the Department of the Environment. Indeed, Urban Programme funding must be viewed in terms of overall public expenditure support to the inner urban areas which, on analysis, reveals far less political commitment than might at first appear. Despite giving apparent priority to the Partnership Areas for example, research indicates that, outside London, reductions in block grant support more than offset Urban Programme funding (Hegarty, 1988).

Funding inner city initiatives in the Partnership Areas may also be adversely affected by the introduction of the Poll Tax, a trend which lends credence to the view that, '. . . in the drive to regenerate urban areas, there is one potential partner which is to play an increasingly minor role – local government.' (Hegarty, 1988).

Regional Development Grants and Regional Selective Assistance

During the financial year 1987/88, the Department of Trade and Industry made available Regional Development Grants, in accordance with regulations introduced in November, 1984 (Figure 9.3); that is, any company registered in the United Kingdom and located in a Development Area could seek project approval and subsequently receive grant support in respect of capital expended on providing new assets, including plant, machinery, buildings or works, or in respect of the number of new jobs created by that expenditure. Qualifying activities included manufacturing processes in Divisions 2–4 of the 1980 Standard Industrial Classification and certain service industries, including banking, insurance and other financial services.

The project under consideration had to create or expand capacity to produce goods or provide services, or result in a material change in the product or service or its process of production. Following project approval, grant aid was automatic and calculated as the higher of either 15% of

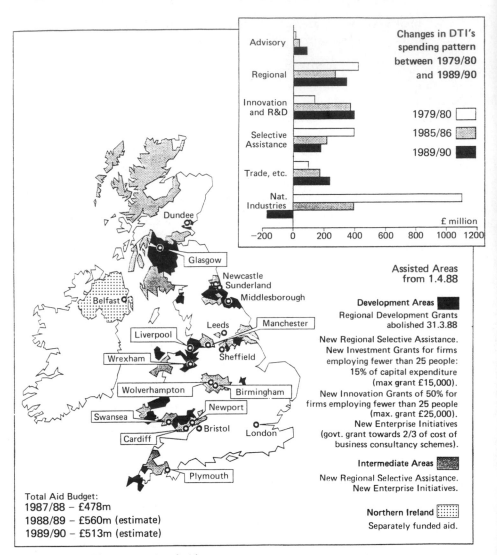

Figure 9.3 The new regional aid map.
Source: The Independent, 13th January, 1988.

eligible capital expenditure, or £3000 for each new job created, net of jobs lost elsewhere in the assisted areas as a direct result of the project.

Eligible capital expenditure concerning the provision of buildings and works included the costs of site preparation and construction, the cost of purchasing a new building not previously occupied and associated professional fees. Adaptation of an existing building also qualified where substantial structural work was involved but the cost of purchasing land was specifically excluded.

The amount of grant, however, may have been subject to limitations, namely;

1. The grant per job limit where support to firms employing more than 200 full-time employees was limited to £10 000 for each net new job created. This constraint did not apply to smaller companies, unless capital expenditure on the project exceeded £500 000, in which case the grant per job limit or £75 000 applied, whichever was the higher.
2. For manufacturing and service projects undertaken by manufacturers, grant payable was limited to 40% of initial capital investment.

In addition to Regional Development Grants, discretionary Regional Selective Assistance, provided under Section 7 of the Industrial Development Act, 1982, were also available during 1987/88, to encourage sound industrial projects which fulfilled the following criteria:

(a) Viability – the project should have good prospects of achieving viability within five years
(b) Need – the applicant must demonstrate that assistance is needed if the project is to proceed
(c) Employment – the project must have beneficial employment repercussions
(d) Efficiency – the project must contribute to the strengthening of the regional and national economy
(e) Private funding – the applicant must generate the greater part of the necessary finance him/herself.

There are two main forms of RSA, training grants and, more particularly, project grants, the latter being related to either the fixed capital project costs or the number of jobs created, normally within three years of its start. The amount of assistance is negotiable and will be the minimum necessary for the project to proceed. Capital related grants are usually paid out in instalments dependent on the company's financial input and progress on the project. Eligible costs include land purchase, site preparation and construction, plant and machinery, and professional fees, but unlike Regional Development Grant, Regional Selective Assistance is entirely discretionary.

Although promotional agencies in the regions relied heavily on the availability of Regional Development Grant to attract new investment, automatic grant aid having a particular impact in providing certainty to investors and, on occasions, reaping substantial returns such as the establishment of the Nissan factory at Sunderland, serious doubts did arise concerning the implementation of RDGs and, in particular, their susceptibility to abuse. The National Audit Office reported that the post-1984 system of grants was more vulnerable to fraud, because of the introduction of job-based grants, especially to service industries, where several prosecutions had resulted (NAO, 1988), while many critics within and outside Government were alarmed at the exploitation of grant aid by large companies, who would have pursued the same investment programmes with or without support.

Substantial revisions, therefore, were contained in the White Paper, 'DTI – the department for enterprise' (Cm 278), published in January 1988. In particular, no new applications for RDG were accepted after March 31st 1988, a deadline which prompted an enormous rush of projects seeking automatic grant aid, as astute companies reviewed their investment plans. In Wales for example, 1300 RDG applications were received by the Welsh Office following the announcement, which compared with a total number of 1640 throughout the whole of the previous financial year. The inevitable delays in processing these claims and the two year period which recipients have to invest their grant, means that Regional Development Grant will continue to be implemented until 1990.

Major projects – Section 8, Industrial Development Act, 1982

Perhaps the least known Government financial incentive available to industrial and business development schemes, concern those projects which are viewed as being in the national interest and which then receive support under section 8 of the Industrial Development Act, 1982. Major capital investment projects involving funding exceeding £0.5m may qualify for grant aid but the selection criteria are stringent and only seven proposals were accepted during 1987. Following the policy review (Jan. 1988) previously referred to, such grants were retitled, 'Assistance for Exceptional Projects' and are now open to both investment and research and development projects.

Although financial support is rare, any firm can apply. Eligible costs are mainly those for fixed capital investment in buildings, plant and machinery, but this expenditure must be likely to provide exceptional benefits to the national economy and substantially enhance the company's competitiveness and productivity. It will normally involve major new products and technical innovation and should have a marked effect on exports from the United Kingdom or import substitution. In addition, as well as demonstrating commercial viability, the applicant must show that without support, the project may be undertaken outside the United Kingdom or that limited company resources would prevent implementation.

The amount of assistance is negotiated on a case by case basis, following a detailed and thorough appraisal, and is the minimum necessary to generate the additional benefits associated with the project. Assistance is usually in the form of grant aid and is taxable as a trading receipt. There is no maximum or minimum level of support but recent awards have averaged 10% of project costs.

Land registers

Registers of unused and underused land owned by public bodies (similar privately held land is excluded) were established under the Local Government, Planning and Land Act, 1980, in order to identify and facilitate the release of sites to the development industry, which otherwise would remain unavailable. The full complement of land registers have been in existence since November 1982 and contribute towards first, pursuing the Government's economic objective of reducing the Public Sector Borrowing Requirement (PSBR) by realizing public assets and secondly, albeit on a small scale, to meeting the constant demands of property developers and especially the volume housebuilders for more land, while easing pressure on the Green Belt.

Registrable sites must exceed one acre and fail, in the view of the Secretary of State, to meet any operational criteria of the landowning authority. By the end of 1987, 66000 acres of land had been deregistered, over half of which had been sold either voluntarily or as a result of ministerial direction, and a further 91000 acres remained available on approximately 9000 sites. Particular powers of enforcement are contained in section 98 of the 1980 Act, whereby the Secretary of State can direct the owner to dispose of specific sites, under prescribed terms and conditions, unless, in response to a notice served under section 99, the authority can show good reason for not doing so. These directions have been regularly utilized.

Considerable emphasis has been placed by Government on its efforts to realize public sector landholdings, in part to deflect Green Belt intrusions but also as a means of promoting inner city renewal. But the development industry has consistently argued that such sites are invariably expensive to develop and often inappropriately located in terms of demand. Overall, their contribution is seen as minimal in the on-going debate concerning land supply, while a recent report by the Audit Commission (Audit Comm., 1988) dismissed the impact of registers as insignificant.

9.4 Recent amendments to grants and incentives

The election of Mrs Thatcher's administration for a third term in 1987, confirmed the continuation of the same broad themes, identified earlier, that have underpinned the Government's financial incentives to industrial and business space developments since 1979; that is, increasingly focusing support on more tightly defined designated areas and specific sites, further diminishing the role of local authorities as vehicles for grant aid and relying more heavily on private funding in order to promote urban renewal. But, the stark political geography that characterized the 1987 General Election result, reflecting the lack of electoral advances made by the Conservatives into Labour's urban heartlands, prompted Mrs Thatcher's election night commitment, to do something for 'our people' in the inner cities. Concentrating primarily on Labour-controlled councils, accused of driving out enterprise through high spending, high rates and a hostile attitude to business, despite the fact that countless municipal initiatives have promoted public/private sector partnerships ranging from joint development companies to industrial investment banks, Mrs Thatcher attacked 'municipal socialism' at the Conservative Party Conference in October, 1987 and outlined a series of measures from the creation of more urban development corporations to educational and housing reform, which '. . . taken together, will greatly reduce the power of local councils . . .' and heralded, amongst other issues, major changes to the system of grants and incentives to industrial and business space developments.

Formal announcement of the amendments, delayed as a result of inter-ministerial power struggles, eventually were contained in the Department of Trade and Industry's White Paper, 'DTI – the department for enterprise' (Jan. 88) and the Department of the Environment's brochure, 'Action for Cities' (Mar. 88), which together establish the grant aid structure to be applied to new proposals in the financial year 1988/89.

DTI – the department for enterprise

The DTI White Paper not only presented a basic restructuring of the department's internal organization into market oriented divisions, but introduced a fundamental shift in trade and industry policy, ushering in profound changes to regional aid, competition and mergers policy and support for innovative projects. In particular, in order to promote the 'Enterprise Initiative', funding and organization have been redirected away from large-scale industrial subsidies and towards the provision of more advisory services, aimed at creating the right climate for initiative and entrepreneurial activity. This administrative and policy reorientation is most obviously demonstrated in resource provision in which the DTI total budget of £2.2bn in 1979/80 will be reduced to £1.35bn in 1989/90. Funds have been redistributed from nationalized industries, which have been privatized or returned to profit, and diverted from regional aid and selective assistance to trade, innovative and advisory services.

Regional Development Grants, the cornerstone of Regional Aid policy for 15 years, were abolished from April 1st 1988 and, with them, the certainty provided by automatic grant aid for projects that met the prescribed criteria. Instead, the White Paper claims that Government will channel more resources to the assisted areas (which geographically remain unchanged) via Regional Selective Assistance, but only where companies can convince the DTI that the planned investment would not go ahead without Government support. In addition, new grant schemes will entitle small companies, located in assisted areas and employing less than 25 people, to receive 15%

of capital expenditure on new projects, up to a maximum of £15 000 and exceptionally, innovation grants of 50% of project costs up to a maximum of £25 000. Funding for these schemes will total £16m in 1988/89 and £48m in 1989/90.

According to the DTI, the overall allocation of resources to regional aid, following these changes, will be an increased budget, from £478m in 1987/88 to £560m in 1988/89, then falling to £513m in 1989/90. However, these figures include existing commitments to Regional Development Grant which will account for £279m in 1988/89 and £209m in 1989/90.

The modified system of regional aid is likely to result in a major shift in aid entitlement from large UK companies, which it is believed, would proceed with new factory development irrespective of inducements, and towards small firms and overseas investors, an anticipated trend which received a cautious welcome from the Confederation of British Industry and the business community at large but has been greeted with dismay by many regional authorities who, while acknowledging the difficulties associated with Regional Development Grants, believe that the lack of guaranteed support in marked contrast to other European countries, will substantially undermine their efforts to promote the depressed regions.

Further concern has been expressed at the level of involvement of and discretion given to civil servants inexperienced in evaluating commercial decisions, but the most severe criticism focuses on the allocation of resources, which are widely regarded as representing a concealed reduction in expenditure. Public rebuke by no less than three previous Conservative ministers at the DTI, Leon Brittan, Sir Giles Shaw and Norman Tebbit, lends credence to the view that a significant proportion of regional funding will be diverted to new marketing initiatives and that substantially less will be available in the form of grants and incentives to industrial and business space developments, thereby restoring tight Treasury control over regional aid.

Action for cities

The long awaited Government attack on the inner cities was finally unveiled in March 1988, in the form of a brochure 'Action for Cities', the contents of which, conspicuously, did not justify a White Paper but, in essence, amounted to a repackaging of existing policies. The statement did however contain important changes to grants and incentives available to industrial and business space schemes which do justify close scrutiny.

Traditional measures which contribute to the Government's approach to urban aid policy, including the Urban Programme and Inner City Partnerships, will continue without modification but the announcement did identify the following initiatives of relevance to the property market's role in urban renewal;

1. A new Urban Development Corporation for Sheffield, located in the Lower Don Valley and covering an area of approximately 2000 acres, where steelworks closures have left a legacy of widespread dereliction
2. The extension of the area covered by the Merseyside Development Corporation to include an additional 800 acres of land
3. The establishment of two new City Action Teams in Leeds and Nottingham, bringing the total to eight, to co-ordinate Government support
4. Additional efforts to bring unused and underused land on to the market by requiring the publication of information about land in public ownership
5. The provision of more managed workshops to be administered jointly by the public and private sectors and developed primarily by English Estates, the Government's factory building and land development agency
6. The introduction of a new City Grant (detailed below) to support private investment in the inner city.

In unison with the Government's publication of 'Action for cities', three private sector initiatives were also announced, namely that;

(a) Eleven major building contractors have pooled resources amounting to £55m and formed British Urban Development to invest in inner city projects
(b) Business in the Community (BIC) is to establish eight teams of business leaders to promote private sector involvement in education, training and investment in urban areas
(c) Investors in Industry (3i), the venture capital group, is to expand its venture fund with a new inner city investment programme.

City Grant. This grant introduced on May 3rd 1988, replaces private sector Derelict Land Grant (DLG), Urban Development Grant (UDG) and Urban Regeneration Grant (URG), all of which, as previously detailed, entailed providing financial aid to urban renewal projects, which would otherwise fail to achieve commercial viability. For, despite the improved links forged between local councils and the business community, stemming from the operation of the UDG scheme and noted in the DoE appraisal carried out by Aston University (Aston Univ., 1988) and the very short life of URG, under which only five projects received support, criticisms of this plethora of essentially similar forms of aid had been expressed.

The process of appraising projects was complex, detailed and protracted and prompted Don Richardson, whose development company was the first recipient of URG to comment that, '. . . they might as well have given me a carrot and a pair of handcuffs' (Wray, 1988), criticisms which inevitably dissuaded many developers from entering into urban renewal projects and where, as a result, private sector DLG, for example, was underspent. This, together with many local authorities simply not having the resources to fund their 25% input to UDG and the Government's commitment to bypassing their involvement anyway, manifest in the URG process, presented the rationale for simplifying the grant structure.

The new City Grant has been modelled on the URG in that it is only available from central government and does not involve local authorities and is aimed at subsidizing the costs of reclamation, refurbishment and new-build projects on a large scale, that is with a total project value in excess of £200000 (indeed, a threshold of £400000 was initially considered). The project, undertaken by the private sector in England (different arrangements exist in England and Wales) must be located in, or have a direct impact upon an Urban Programme priority area (that is, one of the 57 districts required to prepare an Inner Area Programme) and provide jobs, housing or other community benefits.

In making his presentation to the Department of the Environment (or Urban Development Corporation where applicable), the applicant must demonstrate that the project is non-profit making without support, though the proposal will then be subject to the same appraisal process as operated for URG, despite the criticisms of the development lobby (Inner City Directorate, 1988). Funding, on approval, will comprise a grant and/or loan and subject to clawback conditions should excessive profits arise. The first approval under the new City Grant was awarded, not surprisingly in view of the local authority's track record in pursuing grant aid, to a project involving converting derelict buildings in Nottingham.

Overall, 'Action for cities' was disappointing, amounting to expensive packaging but little substance, although the simplification of the grant system was welcomed by developers and professional bodies. As a contribution to urban renewal, however, the statement largely reiterated and extended the broad themes characterizing grants and incentives to developers, identified earlier.

In particular, despite the Government-commissioned study of the UDG in operation, drawing attention to the crucial role of local authorities in identifying local needs and generating appropriate urban renewal projects, the increasing reliance placed on centrally controlled quangos and the cultivation of a more direct relationship between developers and the Department of the Environment, has inevitably called into question the wisdom of increased centralization, rather than genuine devolution of power. This approach places an unnecessary administrative burden on detached central government departments, which prompted the comment that, 'the DoE faces a much greater workload with few resources' (Russell, 1988), while causing dismay at the failure to recognize the major contributions made by many local authorities and epitomized in Sheffield, where the authority had already co-operated with the private sector in formulating schemes to revitalize the Lower Don Valley, only to have an urban development corporation foisted upon it. Not only does this risk alienating local government but is likely to lead to further duplication and wasted effort. The irony of Salford City Council's Salford Quays project appearing on the front cover of 'Action for cities', was not lost on disillusioned local authorities.

Further severe doubts concern the role of private investment, which is inherently demand-led and appears unlikely to be sufficient to tackle problems on the scale and in the locations required. The necessity for substantial public sector financial inducements is unquestionable but, in fact, urban aid funding, on Mrs Thatcher's own admission, includes no new money and in practice, is countered by expenditure cutbacks in related programme areas.

The strategic policy approach to the inner cities continues to be characterized by ad hoc decision making and a complicated assortment of grants and incentives, which lack any properly integrated planning framework within which co-ordinated development initiatives can operate efficiently.

9.5 Conclusion

There is little doubt that, despite the deteriorating outlook for the balance of payments, the recent more favourable economic indicators including the improvement in manufacturing output and the fall in the official unemployment total from its peak of over 3 200 000 in July 1986, do represent an improvement in the overall economic well-being of the majority of people. It is equally undeniable that severe regional economic imbalances continue to exist and that the significant minority who have not benefited from the economic upturn, are disproportionately represented in the inner cities. The Government's own regional statistics (HMSO, 1988) continue to confirm out-migration from the northern regions, Scotland and the West Midlands and an increasing discrepancy in regional gross domestic product and personal disposable income, while new employment opportunities continue strongly to favour the southern regions (MacInnes, 1988).

Such economic discrimination between regions and within cities demands a radical reshaping of Government policies, aimed at tackling private capital's reluctance to invest in these areas. This can only occur when Government recognizes the significance of regional infrastructure provision, the role of public sector agencies, including local authorities, in committing sufficient public investment to attract private capital, its own role in encouraging for example the creation of regional venture capital funds in collaboration with the investment institutions and, above all, by formulating a positive national and regional planning and development framework, in which regional and urban renewal objectives are fundamental.

In practice, this recognition has not occurred and Government continues to pursue an incremental approach to supporting and promoting industrial and business space developments, an approach which invariably leaves industrialists, investors and developers bemused and, all too often, assessing alternative proposals elsewhere. Grants and incentives to developers are fewer, more difficult to obtain and usually involve protracted negotiations, a combination which rarely provides the sought after certainty and thus financial confidence necessary to convert marginal or loss-making projects into viable propositions. This half-hearted commitment to regional and urban renewal does not bode well for the future but, there are more optimistic signs and, ironically, they emanate from the Conservative Party's own supporters. In pursuit of electoral advances, the Tory Reform Group for example, advocate an assertive and co-ordinated regional development policy to promote the recovery of the Party in the northern regions, while the Bow Group support the more radical approach of weighting taxation in favour of the regions in order to stem the outflow of skills and capital. Most recently, the SANE group of Tory backbench members of parliament, concerned at the environmental impact of over-centralized development in the south, favour extending Government intervention, reviewing infrastructure investment and public sector employment plans and utilizing public investment and tax incentives to attract business and industry away from the south, all components of an explicit national planning and development strategy. At present, these policies contrast conspicuously with the market-oriented ideology of the Government but, as development pressures come into increasing conflict with more conservation oriented local plans in the south, the Government may find that a further more meaningful restructuring of grants and incentives to promote industrial and business space developments in outlying regions, becomes a political necessity.

References

Armstrong, H. and Riley, D. (1987) The north–south controversy and Britain's regional problems. *Planning Outlook*, 30 January.

Audit Commission (AC) (1988) *Local Authority Property – an Overview*, HMSO, London.

Bow Group (1988) *Regional Policy*. Bow Group.

Brittan, L. (1988) View from the back benches. *Town and Country Planning*. February.

Central Statistical Office (1988) *Regional Trends 23*. HMSO, London.

Comptroller and Auditor General, (CAG) (1988) *DoE Derelict Land Grant*, HMSO, London.

Dennis, R. (1978) The decline of manufacturing employment in London. *Urban Studies*, **15**, 63.

Department of Land Economy (DLE) (1987) *South-east Employment and Housing,* Cambridge University.

DoE (1988) *Action for Cities,* HMSO, London.

DoE/Aston University (1988) *Evaluation of the Urban Development Grant,* HMSO, London.

DoE Inner City Directorate (1988) *City Grant Guidance Notes,* HMSO, London.

Economic and Social Research Council (ESRC) (1985) *Changing Cities,* ESRC Inner City Research Programme, September.

Hall, P. (1988) The Industrial Revolution in reverse. *Planner,* January.

Hegarty, S. (1988) Inner city aid; the figures behind the hype. *Public Finance and Accountancy,* 9 January.

Institute for Employment Research (1987) Local Prosperity and the North–South Divide, Warwick University.

MacInnes, J. (1988) *The North–South Divide; Employment Change in Britain,* CURS Research no. 34, University of Glasgow.

National Audit Office (NAO) (1988) *Arrangements for Regional Industrial Incentives,* HMSO, London.

Pearce, G. (1988) Lessons from the Urban Development Grant Programme. *Planner,* April.

Robson, B. (1988) Regional futures; the good and bad omens. *Town and Country Planning,* February.

Russell, A. (1988) One man's view. Action for cities? *Estates Gazette,* 4 June.

Taylor, R. (1988) North–south. *Planner,* February.

Wray, I. (1988) Worth its weight in red tape *Architects Journal,* 25 May.

10

The role of development control in implementing industrial and business space developments

10.1 Introduction

By its very nature as the principal means of implementing the policy content of statutory development plans, the planning control system, exercised primarily by local planning authorities and/or the Secretary of State at the Department of the Environment, is a crucial facet of the industrial and business space development process. Indeed, the public control of land use and development, applied via our elected representatives, is a fundamental element in the British model of town and country planning.

The term 'development control' is used to describe the decision-making activity in which planning applications to develop or change the use of land and/or buildings are processed and includes any subsequent challenges at appeal. But, the degree to which development plans are implemented by the planning control machine and the level of interpretative discretion involved in the consideration of particular planning applications, depends on a variety of factors, not least the political stance of the authority and the expectations of the local electorate. Furthermore, and debatably of greater significance since 1979, is the issue of compatibility between planning policies and market forces and, where planning objectives and market demands conflict, the reaction of the Secretary of State in establishing national policy and implementing that policy in particular, at appeal. These development control characteristics are especially important when considering industrial and business space developments because, in the light of the severe economic recession in the early 1980s, the relationship between planning control and industry has been focused on by the Government as a contributory factor to the slump in economic activity and, therefore, an administrative area ripe for reform.

This is not to suggest that reformist zeal began in 1979. Development control has been subject to constant political and professional debate since

1947 and periodically formally examined by central government (see for example, Dobry's Review of the Development Control System, Final Report [DoE and Welsh Office, HMSO, 1975]). Nevertheless, although the scope and nature of development plans have broadened since 1947 and the Government have regularly enacted minor amendments to, for example, the definition of development and, thus, the need for planning permission, the fundamental characteristics of planning control remain. The system continues to be essentially regulatory (with a degree of discretion in its interpretation of development plans and supplementary guidance) and reactive to developer initiatives, relying primarily on moderating and channelling development proposals rather than actively generating them. In some circumstances, notably areas under considerable development pressure, the development control machinery has been strengthened, as market demands generate an increasingly hostile local political reaction (areas ironically, where local planning authorities are often Conservative controlled). Elsewhere however, since 1979, the Government have introduced a number of initiatives and modifications aimed at relaxing and reducing controls, especially those affecting industrial and business space developments.

10.2 The relaxation of development control

The introduction of Enterprise Zones and Urban Development Corporations, via the Local Government, Planning and Land Act, 1980 (Chapter 9), was an early demonstration of the Government's intention to restrict or remove administrative and financial burdens on business, while the DoE White Paper, 'Lifting the Burden' (HMSO, 1985) included further proposals on the same theme, such as the introduction of Simplified Planning Zones and the review of the Use Classes Order (the impact of which is discussed in section 10.4).

In particular, the Government's efforts to introduce an enterprise culture, removing the obstacles to economic activity, especially, in the form of new businesses, has prompted a series of circulars, aimed at influencing the development control stance of local authorities, when planning applications for industrial and business space developments are under consideration. DoE Circulars 22/80, 16/84, 2/86 and 13/87 (see References), summarized in Planning Policy Guidance 4 (HMSO, Jan. 1988) entitled 'Industrial and Commercial Development and Small Firms', combine to present the Government's planning position regarding industry and business space. In pursuit of planning certainty, the circulars urge local planning authorities to:

1. Respond 'constructively' and promptly to enquiries and applications
2. Give priority to applications which contribute to national and local economic activity

3. Only oppose development where this serves a clear and justifiable planning purpose and the economic effects have been considered.

In order to fulfil these goals, the circulars emphasize;

(a) The need to promote a creative climate towards industrial and business schemes by adopting a flexible approach to interpreting development plans and ensuring that sufficient suitable land is available
(b) The changing nature of industry and business, much of which is now compatible with other uses, notably residential, from which non-conforming industrial uses were often weeded out in the past; rigid zoning is now necessary only in exceptional circumstances
(c) The need for realism in determining planning and development strategies, that is, an awareness of market viability especially where this entails the reuse of redundant buildings
(d) The importance of hi-tech and mixed-use projects to industrial regeneration
(a) A relaxed approach to enforcement proceedings should this jeopardize industrial and business activity
(f) The Government's opposition to unnecessary constraints on market forces, such as local user conditions and their belief in the market's ability to assess needs.

Clearly, such advice when applied by local planning authorities to particular applications, will be interpreted with varying degrees of discretion in response to local economic variations, but that discretion is coloured by the knowledge that the applicant has the right of appeal, that the number of appeals has risen substantially since 1979 and that the success rate has increased too. This is perhaps best illustrated by the loss of industrial land and buildings to retail warehouse users.

The Government's rationale behind their approach is well known. The planning system and the development control process in particular, are accused of perpetuating inflexible and time-consuming procedures which undermine the financial viability of individual schemes and which, in the case of industrial and commercial projects, affect production, employment and, thus, economic growth. Delays and economic losses are caused by development plans that are dated and vague, overly detailed consideration of individual applications, protracted administration and, in particular, an inadequate understanding of market forces. Such charges carry some credibility but take no account, for example, of delays caused by ill-prepared planning applications, an all too common occurrence. Similarly, market fluctuations and their impact on funding provision, often explain failure to act on a planning consent but the planning system itself is a soft target, a convenient political scapegoat for poor industrial performance and, as a result, the subject of extensive tinkering.

With regard to industrial and business space developments, the introduction of Simplified Planning Zones and the review of the Use Classes Order, are probably the most significant changes and deserve special attention. In addition however, it is important to stress the increasing emphasis being placed on administrative discretion and, in particular, the negotiated solution as a means of facilitating the industrial and business space development process and the role of business elements in mixed-use projects, which may be implemented as planning gains.

10.3 Simplified planning zones

On May 1st 1984, in a speech to the Royal Institution of Chartered Surveyors, the then Secretary of State at the Department of the Environment, Patrick Jenkin, announced the imminent publication of a consultation pape on Simplified Planning Zones, which, in utilizing the Enterprise Zone model, . . . would give local authorities powers to take the initiative in granting planning permission for development in parts of their areas, without the need for a planning application . . ., and therefore facilitate development.

The Department of Trade and Industry Report, 'Burdens on business' (HMSO, 1985), focused the use of such zones on industrial and business space users in particular and the subsequent White Paper, 'Lifting the burden,' further emphasized their role. Enabling powers were contained in Part II of the Housing and Planning Act, 1986, and the Town and Country Planning (Simplified Planning Zone) Regulations, 1987, (SI 1987 no. 1750), finally came into force on November 2nd 1987, together with Circular 25/87 which specified their purposes and procedural requirements.

In meeting the general objective of confining conditions and limitations on development to the minimum necessary to maintain health and safety standards and to safeguard the environment of surrounding areas, Simplified Planning Zones are intended to provide a further means of improving the efficiency and effectiveness of the planning control system and thereby, stimulate the development process. By discouraging delay and disincentives to development, SPZs are seen as particularly appropriate in older urban areas in need of revitalized economic activity, where their role may complement other development incentives, such as advice on the availability of sites and/or financial assistance and thus form part of a comprehensive package of measures to promote urban regeneration, capable of being actively marketed. This orientation of the SPZ concept explains their specific exclusion from National Parks, Conservation Areas, Areas of Outstanding Natural Beauty and Green Belts.

A Simplified Planning Zone is an area in which an SPZ scheme is in force. Where a scheme has been formally designated, planning permission for specified development is deemed to be granted, an approach comparable with a localized General Development Order. That consent may be subject to exceptions – (for example, Classes B3 – B7, Special Industrial Uses, identified in the Use Classes Order, 1987, are not viewed as appropriate for inclusion in an SPZ). However, such provisions should not undermine the basic principle of SPZs, namely to minimize the impact of planning control where possible. It should be emphasized, however, that deemed planning consent does not negate the need for other permissions including, for instance, listed building consent and Building Regulation approval, as noted in DoE Circular 25/87.

Every local planning authority is obliged to consider the desirability of making an SPZ scheme and procedural requirements are determined accordingly but, unusually, any other body or individual, including landowners and developers with strong vested interests, can request a planning authority to implement an SPZ and, where refused or ignored, can require the matter to be referred to the Secretary of State for his decision. Despite this safeguard, however, the process of approval remains considerable and is presented in Figure 10.1. The proposals contained in the scheme must be adequately publicized and open to representations and a local enquiry will consider objections and may result in modifications, prior to the authority's final adoption by council resolution. Copies of the scheme must be sent to the Secretary of State, who has the power to require the scheme to be submitted to him for approval. Whether adopted or approved, the Simplified Planning Zone has a lifespan of ten years from the date of designation.

Although Simplified Planning Zones are not restricted in size or required to concentrate upon particular uses, they were perceived as being especially applicable to industrial and business space uses and might well focus on large derelict industrial sites but were likely to amount to hundreds rather than thousands of acres. Early practice has confirmed these expectations.

The SPZ scheme consists of a map, a written statement and appropriate illustrative material. In determining the nature of the deemed planning permission, the scheme will be either specific, in which the particular types of permitted development will be itemized, together with any limitations, or general, where the consent amounts to a blanket permission covering most types of development subject to exceptions, the latter being precisely defined. Although the provisions of the scheme should avoid detailed facets of developments, the SPZ must maintain health and safety standards with regard to, say, pollution and highway design and may also determine a suitable form and scale of construction in the light of surrounding environmental factors. Density, height of buildings, parking and landscaping may all be subject to policy, while sub-zones within an SPZ may also be incorporated, for example, to meet the needs of areas adjacent to hazardous users.

The DoE Planning Policy Guidance Note (PPG5, Jan.1988) on Simplified Planning Zones draws attention to industrial and business space development situations compatible with SPZ designation and, in practice, initial

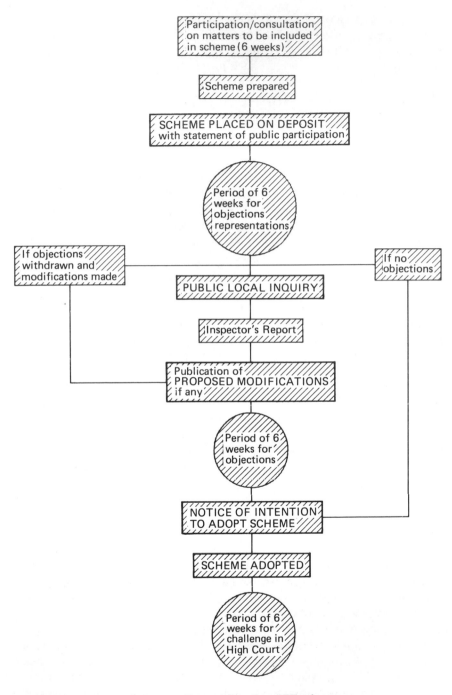

Figure 10.1 Preparation of an SPZ scheme.
Source: PPG note 5 (DoE 1988)

interest from local authorities has been focused accordingly.

Corby District Council, one of the pioneering Enterprise Zone authorities, have pursued the SPZ concept with enthusiasm, in particular with regard to the former steelworks sites of 400 acres, where general industrial and warehousing uses will predominate. Local debate concerning the suitability of the SPZ designation mirrors national reaction to the introduction of the concept and involves evaluating the following costs and benefits. Advantages include:

1. A reduced number of planning applications and appeals will improve efficiency and save time, cost and manpower resources
2. A more positive form of planning control which would provide much sought after certainty to developers and investors, as well as relief from planning application fees, which for full permission on non-residential buildings, currently costs £66 per 75 sq.m of gross floor space up to a maximum of £3300 for 3750 sq.m or more
3. An opportunity to use the SPZ designation as a promotional tool to stimulate development, without necessitating the assembly of sites but which, in association with financial and other inducements, would amount to a comprehensive commercial package.

Concern however, has been expressed regarding:

(a) The heavy workload involved in the designation process, which may extend rather than reduce bureaucracy and generate more public confusion and uncertainty for developers instead of removing unnecessary administrative hurdles
(b) The loss of planning application fee income which in the case of Corby District Council, was estimated at £30000 per year
(c) The apparent inflexibility of a system based on fixed planning and development criteria and the environmental repercussions
(d) The loss of the right to object to specific developments and the loss of control over, for example, design standards which may present neighbouring users with an inferior environment.

It is clear, that a limited number of local planning authorities will utilize the SPZ concept, especially those experienced with similar approaches and that industrial and business space developments will emanate from such designations. The level of local commitment to Simplified Planning Zones is likely, however, to fall short of Government expectations, not least because many authorities will argue that without the necessary financial inducements to investors (aid, which as noted in Chapter 9, has become more difficult to obtain), administrative streamlining alone will not attract investment and where demand for industrial and business space development exists, Simplified Planning Zones will be an unnecessary administrative luxury. In the case of Corby however, SPZ status was confirmed.

10.4 The Town and Country Planning (Use Classes) Order, 1987

The Town and Country Planning (Use Classes) Order, 1987, came into force on June 1st 1987, updating and simplifying the 1972 Order, while DoE Circular 13/87, Changes of Use of Buildings and Other Land, provided further explanatory guidance. The basic rationale for this review, carried out by the Property Advisory Group, stemmed from the emergence of new types of development, such as hi-tech industrial estates and fast food retailers (uses which were not easily classified) and, as previously noted, the Government's commitment to lifting the burden on enterprise and economic initiative.

First introduced in 1948, the Use Classes Order, which applies to England and Wales, details the latter part of the legal definition of development, then contained in the Town and Country Planning Act, 1947, which includes, '. . . the making of any material change in the use of any buildings or other land'.

In particular, the objectives of the new Order are to broaden the scope of use classes where appropriate and thus reduce the number of individual classes and therefore planning applications, aims which have notable repercussions for industrial and business space developments.

Dividing uses into the four broad categories of shopping area, business and industrial, residential and social, and community uses (and codified A,B,C and D), the 1987 Order contains significant changes of relevance to industrial and business space developments. Most notable are:

1. A new business class (B1) which amalgamates many former Class II office uses with former Class III light industrial uses and is most obviously demonstrated in evolving hi-tech users, who combine office, light industrial and research and development activities, such as the manufacture of computer hardware and software, within the same building. This class therefore covers offices and/or any industrial process that can operate in a residential area without causing environmental damage to that area. Clearly, such buildings have considerable user flexibility and are unlikely to resemble traditional light industrial (shed) units but it is important to note that existing light industrial structures may also have an opportunity to change to office use without planning permission. This conversion may contravene traditional local plan policies but the potential increase in rental income may well prove financially appealing.
2. Class A2 – Financial and Professional Services – is aimed at those activities which historically fell between offices and shops and focuses on services which have face-to-face contact with the general public, such as banks, building societies and estate agents and are thus deemed compatible with traditional shopping uses.

Notes

1. In the case of Vickers Armstrong v Central Land Board (1957), it was held that offices attached to a factory are regarded as part of the class in

which the factory falls and the offices have no independent use rights and, therefore, it is the primary use of a building or other land that determines its use class.

2. The new Use Classes Order necessitated a revision of the General Development Order, that is, the Order which specifies certain types of development, including changes of use, with regard to which planning permission is deemed to be granted. In particular, the Town and Country Planning (Amendment) (No. 2) Order, 1987, notes that where floor space does not exceed 235 sq.m. planning consent is deemed granted for changes of use from, (a) Business (B1) or General Industrial (B2) to Storage and Distribution (B8); and (b) General Industrial (B2) or Storage and Distribution (B8) to Business (B1), subject to further planned amendments, detailed below.

Interpretation and implementation

Inevitably local planning authorities are likely to vary in their interpretation of the provisions of the new Order and, although owners and occupiers do have greater scope for change of use, the authorities' enforcement powers remain available to deal with breaches of planning law, in the light of which it is important to determine:

1. To what land and/or buildings does any existing permission refer?
2. Into which Class does the current activity fall under the old Order and the new Order?
3. How does the existing permission describe the use and, in particular, does that permission contain conditions or restrictions on the use? Such conditions are not uncommon and planning permission would be necessary should the applicant require a condition to be relaxed or revoked. However, conditions attached to current and new planning permissions may, on referral to the DoE and following DoE Circular 1/85, be deemed unreasonable and discharged.

The impact of the new Order will undoubtedly vary from area to area, depending primarily on local economic circumstances, the policies contained in the adopted statutory development plans, the planning stance of the authority, comparative rental levels in the locality and lease terms in existing landlord and tenant agreements. It does however appear likely that the new B1 Business Class will prove the most significant amendment. In areas of, for example, high office demand, the ability to change from light industrial to office use without consent could be widely utilized, but it should be noted that such a change is conditional on the office not serving the public directly and that the amenity test is observed. Indeed, it has been suggested that the amenity test will be strictly applied by some authorities as a means of minimizing the effects of the new Business Class.

Where the premises in question are let, potential modification of use within the new Business Class will also be subject to the terms of the existing lease. User and rent review clauses will be especially important. Where for example, a user clause restricts the use to, say, Class III (1972 Order), a tenant would be prevented from taking advantage of the new Business Class in the 1987 Order. Similarly, a landlord could not increase the rent charged although negotiated settlements between landlords and tenants seem likely.

Although the revised Use Classes Order may encourage economic initiative in areas where business use demand exceeds a limited supply of suitable premises, it does not provide property developers with limitless opportunities. Local planning authorities retain most of their development control powers and in areas where, for example, restrictions on office uses have been pursued in order to protect industrial space and local employment opportunities, other means of pursuing this objective may be utilized. Thus, despite the opposition of the Secretary of State, conditional consents may be used to try and minimize the user flexibility contained in the Order, including local user conditions, or, alternatively, restrictive planning agreements under Section 52 of the Town and Country Planning Act, 1971. Following Circulars 1/85 and 13/87, such conditions may be deemed unreasonable by the Secretary of State, but this would necessitate a planning appeal. Nevertheless, effectively implementing traditional planning policies concerning industrial developments, will prove more difficult for planning authorities, especially where latent business user demand exists and these opportunities will be exploited by astute entrepreneurs (Marsh, 1988).

Further amendments and refinements are included in the consolidated version of the General Development Order, 1988 and the explanatory regulations contained in DoE Circular 22/88. The new Order which came into force on December 5th 1988, replaces the 1977 Order and twelve subsequent amendments. Several modifications affect industrial and business use development, in particular:

1. A use falling within Class B2 (General industrial) can now be changed to the B1 Business class, without a limitation on the floor area concerned, an alteration which could have a significant impact in, for example, London Boroughs adjoining the City of London, where one estimate suggests that 16 m sq.ft of industrial property could be converted to offices, without physical alterations requiring planning permission. However, a limitation on floor area still applies to changes of use involving Use Class B8 (Storage and Distribution). Change of use from B8 to B1 and from B1 or B2 to B8 are permitted, but only where the total amount of floor space used for the purposes of the undertaking does not exceed 235 sq.m.

2. Change of use from Class A3, Food and Drink, to A2, Financial and Professional Services and from A2 to shops, become permitted develop-

ment within the 1988 Order and thus, in the former case, further inhibit those authorities committed to controlling the spread of service offices in town centres.

The removal of the power enabling highway authorities to direct local planning authorities on how to decide applications affecting classified roads, in favour of a highway authority right to be consulted before a decision is made, will facilitate industrial and business space schemes which generate significant traffic.

10.5 Negotiated agreements

Historically, the basic relationship between the planning system and market forces was clearly established, in that local planning authorities prepared simple land use plans, which in essence were responsive to market demand and subsequently implemented those plans, primarily through the development control mechanism. This approach provided certainty to the parties concerned and emphasized the regulatory and reactive characteristics of planning control.

Since the mid-1960s, however, development plans have broadened in scope significantly, with varying degrees of achievement, in response to newly emerging issues, notably emanating from a fundamental restructuring of the economy which has not favoured the inner urban areas and the outlying regions. More recently turbulent market conditions, taking place in a political context that is largely dependent on the market to resolve issues and often perceives the planning system as an unnecessary obstacle to development and therefore economic activity, have combined to generate considerable uncertainty in the relationship between the planning system and the market. In such circumstances, it is inevitable that the parties will seek to discuss development possibilities informally before the presentation of planning applications. Indeed, negotiation with developers is an increasingly critical element in the process of development control, not least because it may provide opportunities for the local planning authority to negotiate community benefits, which may well comprise, directly or indirectly, industrial and business space inclusions.

The notion of planning gain or community benefit generates varied reactions among the members and officers of local planning authorities and prompted the DoE to issue Circular 22/83 on the subject, in which planning gain was defined as: '. . . a term which has come to be applied whenever, in connection with a grant of planning permission, a local planning authority seeks to impose on a developer a positive obligation to carry out works not included in the development for which permission has been sought, or to make some payment or confer some right or benefit, in return for permitting development to take place.'

The tone of this definition and its emphasis on the local planning authority imposing its will on the developer, is indicative of the Government's general reaction to planning gain but, as will become clear, amounts in practice to little more than political window dressing.

Negotiations concerning the detailed aspects of development projects, such as density and intensity of use, accessibility, highway, parking and landscaping standards, are of course commonplace, but the essential prerequisites necessary to enable the planning authority to negotiate inclusions in development proposals, is conspicuous market demand for a particular use and a limited supply of suitable sites. Where these conditions exist, the granting of planning consent for, say, retail warehousing on land zoned for general industrial use, a common occurrence especially on appeal, would produce an enormous increase in the value of the land. An opportunity for the community, via their agents, the local planning authority, to appropriate a share of the resulting capital appreciation, may therefore arise. However, the necessary market conditions are not common and, thus, a strong spatial discrimination between planning authorities and their ability to negotiate benefits is a fundamental feature of planning gain. Furthermore, that capability is not fixed, but subject to the dynamics of market forces so, for example, some district authorities located on the fringes of the conurbations, previously of little consequence to developers but now under pressure to allow out-of-town shopping centres, have permitted business workshop units which would otherwise be ignored by the market, because higher yielding development opportunities existed. Similar localized bargaining strength results from major infrastructure developments such as the completion of the M25, the expansion of Stanstead Airport and the construction of the Channel Tunnel.

The geographic discrimination that characterizes planning gain negotiations is obviously inequitable and justifies the abolition of the practice (despite the argument that those who suffer development are entitled to the reward) and its replacement with an enforceable levy on the increase in land value attributable to the receipt of planning consent (betterment), accruing to the national Exchequer and distributed on the basis of need for the provision of community facilities. Our inability to devise such a levy has of course consistently undermined the planning system and '. . . this has proved perhaps the biggest structural failure of British planning in its 40 year history' (Hall, 1988).

That failure has prompted many suggestions including the public purchase of land in advance of development on Swedish lines (Reade, 1987), the taxation of land deemed ripe for development as administered in Denmark (Ambrose, 1986) and the auctioning off of development rights in the form of Development Bonds (Hall, 1988). But with the demise of the easily abused Development Land Tax, which netted a paltry £55m in 1983/84, no direct means of recovering betterment nationally exists, leaving only those favoured authorities, market led into a strong bargaining position, capable of effectively levying a local fiscal equivalent, planning gain, which may be paid in cash and/or kind. Very occasionally, that levy may yield enormous sums. Thus, for example, the London Borough of Tower Hamlets was able to negotiate a package of community benefits, in association with the commercial content of the Spitalfields Market development proposals, worth approximately £20 m, this is an area where the annual Housing Investment Programme allocation amounts to little more than half that sum.

The inequitable characteristics of planning gain and the moral and ethical considerations that the practice evokes are the subject of continuing debate but, whether one views bargaining as an acceptable facet of planning or not, pre-application negotiations between the development control officers (and occasionally, members) of the planning authority and prospective developers have become '. . . an integral part of our planning legislation and cannot be abandoned without reconsidering the whole system of planning' (RTPI, 1982).

The trend towards negotiated solutions

In explaining the problems associated with negotiated inclusions and their impact on industrial and business space developments, it is initially helpful to identify more fully the reasons why the practice has apparently come to the force during the last ten years and thus illuminate the context in which discussions take place. The explanation for the emergence of the practice of bargaining between planning authorities and developers is a combination of historic precedent and recent economic and political circumstances, factors which can be identified as follows:

1. In the United Kingdom, there is extensive experience of partnerships between public and private sector agencies (Chapter 4), in which joint development ventures, often involving very large scale projects such as town centre redevelopments and major urban renewal schemes have featured strongly. The physical, environmental and financial constituents of these undertakings emerged as a negotiated solution, in which the parties defined a mutually acceptable development package, usually utilizing the authorities' land assembly and infrastructures powers, with the developer's expertise, funding provision and entrepreneurial flair.
2. The property boom of the late 1960s and early 1970s again focused media and thus public attention, on the enormous speculative gains that could be achieved on receipt of planning permission. The protracted saga of Centre Point and later Tolmers Square in London (Wates, 1976) emphasized the potential developments. Not only did this prompt the then Labour Government to attempt again, like its post-war predecessors, to harness inflated development values, this time in the form of the Community Land Act, 1975, which, albeit briefly, institutionalized negotiated solutions, but local

authorities too saw an opportunity in which they might achieve clawback and thereby reap both financial and, in particular, political profits. That stance has been underpinned by the widely-held view that unearned increments emanating from development plan zoning and subsequent planning consents, were created primarily by and should be returned to the community, in whole or at least in part.

3. The last ten years has witnessed a rapid escalation in the pace of economic, social and technological change, perhaps most vividly demonstrated recently in the massive upsurge in demand for office accommodation in Central London, with unobstructed floor space exceeding 40 000 sq. ft to meet the requirements of deregulated financial organizations and, in the surge of planning applications for enormous retail facilities containing over 1 m sq.ft of floor space, together with integrated leisure elements, in out-of-town locations. Such developments have undermined the credibility of a supposedly forward-looking but invariably protracted development plan process, which all too often has reacted slowly to these pressures. As a result, the negotiated solution, in the absence of strategic guidance in the metropolitan counties and on occasions involving departures from long established plans or, under threat from potentially successful appeals (the rate of which has increased considerably under the Conservatives), has increasingly been seen as a constructive response to immediate and possibly overwhelming development pressures. Indeed, some authorities have viewed limited departures from adopted plans as a method of defusing those pressures, a reaction that has been facilitated where the developer has included community benefits in his scheme. By setting such precedents, however, they inevitably make similar responses to subsequent proposals difficult to avoid. The impact of such circumstances in the West Midlands recently caused the local Regional Forum to express concern that the region's economic regeneration was being impeded by the lack of serviced industrial land and that, despite an increase in demand for sites, more than 200 acres of high-quality industrial land was lost to other uses, notably retailing. The role of development control in implementing industrial strategies and facilitating industrial and business space developments, is clearly capable of being undermined, where strong demand for alternative uses bolsters hope value.

4. The period since the 1979 General Election has also been characterized by a steadily diminishing political commitment to any serious notion of medium- to long-term planning objectives. Indeed, the planning system and especially development control activity, has often provided a convenient political scapegoat for economic recession, social breakdown and unacceptable environmental conditions. In the light of the Government's continuing market-oriented economic strategy, combined with an ideological stance that views the planning system at best as a means of oiling the development machine, it is hardly surprising that planning has become increasingly

centralized, budget-related and incremental in character. In that climate of ad hoc decision-making, any method of implementing planning policy will be attractive and, if that method should bypass Central Government restrictions on expenditure too, then a spirit of compromise may prevail.

5. The property development industry is notoriously vulnerable to booms and slumps in activity levels, in part self-inflicted by a propensity to oversupply the latest fashionable development, a characteristic most recently demonstrated in hi-tech industrial schemes and potentially with out-of-town shopping complexes. Delays, including fighting planning appeals, are not only expensive but do not assure the developer of success and, even if fruitful, market conditions may have changed. The negotiated solution will minimize unnecessary delay (although critics often consider delay as a tactical negotiating tool) and generate certainty and, as such, will therefore reduce development risk and development costs. Of equal significance, the solution may also prove beneficial to the developer in that a more lucrative use, or intensity of use, may be incorporated into the scheme by proffering community benefits than would otherwise have proved acceptable to the authority, in which case the developer may achieve an immediate speculative gain on his investment. In addition, as will become clear, the developer has inherent advantages in the negotiation process which, if exploited adroitly, will ensure more than the minimum return on his investment, necessary to go ahead. That likelihood may in part explain the half-hearted restrictions on planning gain contained in Circular 22/83 which appear to leave the negotiation door open!

Overall, it is clear that, if the present economic and political decision making context persists, the product of negotiated solutions will continue to reflect the bargaining strengths of the respective parties. In many cases that amounts to a weakened planning system (in part perpetrated by the Government itself) combined with low levels of demand which together generate few if any opportunities for the authority to negotiate inclusions. But, in growth areas, buoyant market demand, which may also conflict with strongly defensive planning policies, especially in the inner London boroughs and the Conservative controlled shire counties and districts may place the authority in a position to negotiate inclusions in selected development schemes and those requirements can include industrial and business space elements, where appropriate.

Planning gains in practice

Community benefits included in development schemes as a result of negotiations between local planning authorities and developers are extremely varied, both in nature and scale. With particular regard to industrial and business space developments, examples include the following:

1. Specification of use(s) is probably the most common instance of planning gain in practice, in which, for example, a residential element is included in a predominantly commercial proposal, or light industrial or business space units are aligned with retail warehouses. In effect, highly profitable uses subsidize otherwise marginal components and, in some cases, negotiations have developed further to include reference to the identity of the user as well as the use; for example, the local housing association may receive user nomination rights or workshop units may be handed over to a community trust who subsequently manage the space in accordance with local needs.

2. Dedication of land and buildings to public use. Land made available as public open space or for road improvement purposes often evolves from negotiated solutions, while larger commercial schemes may incorporate more substantial benefits, such as the construction of community or training centres and the funding of community programmes, libraries or sports facilities or the restoration of derelict buildings, including former industrial premises, now incompatible with contemporary demands.

3. The provision of infrastructure. Although requirements regarding infra-structure provision are mandatory in some countries, it is only in the light of cutbacks in public expenditure in the United Kingdom that contributions by the developer towards off-site infrastructure costs have become common. These inputs by developers arise especially where the project is acceptable in principle but would be refused on the grounds of prematurity, because inadequate services are subject to financial constraints. Similarly, where the project would generate traffic congestion on local roads, the developer may offer to expand their capacity, not only to counter planning objections but to facilitate accessibility to the scheme and thus reduce development risk.

4. Cash payments are acceptable in certain circumstances (identified in DoE Circular 22/83) in lieu of more concrete inclusions. In addition, despite general and especially media condemnation of cash payments and, in particular, off-site gains unrelated to the site in question, such arrangements are far from rare and invariably follow the developer's initiative. Bearing in mind that the developer's essential objective in pursuing development opportunities is to maximize profits and the return on his investment, suggestions of cash in lieu or off-site gains, stemming from the developer, only arise because they are financially expedient to him. Not only does this approach maximize the high earning commercial content of the main scheme but, more importantly, it facilitates the funding, letting and eventual sale of the project to the investment institutions. In contrast mixed-use schemes, combining say offices and light industrial floor space will be less favoured, in that office rents may be adversely affected by the close proximity of industrial uses, borrowing costs may be higher and, as a result, the anticipated investment yield will reflect greater risk. The provision of cash in lieu or off-site gains may well be viewed as a relatively cheap means of securing a much improved investment.

In the light of such practices, how much credence should be given to the argument that, by negotiating planning gains, local planning authorities, via the development control mechanism, are holding the development industry to ransom and trading in planning permission? The DoE's own Property Advisory Group (made up almost entirely of Chartered Surveyors who presumably reflected the development industry's views) appeared to provide a sympathetic Government with every reason to abolish the practice by commenting that '. . . as soon as a system of accepting public benefits is established which goes beyond the strict consideration of the planning merits of a proposed development, the entire system of development control becomes subtly distorted and may fall into disrepute' (PAG, 1981).

It would not be overly cynical to suggest that the rhetoric of the Property Advisory Group amounted to political window-dressing and that, in truth, the advantages of leaving negotiated solutions largely unconstrained by legal definitions or even procedural controls, favoured the development industry (and for that matter, the Government, who benefit from the provision of community facilities at no direct cost to the Exchequer) far more than the disadvantages associated with such practices. By specifying the benefits accruing to the developer, an indication of how local planning authorities can improve their bargaining position in securing planning gains, including industrial and business space elements, will begin to emerge.

Advantages to the developer in negotiation

It is already clear that the present economic and political climate creates a promising decision-making framework, within which the astute development team (which often includes head-hunted local authority planners) can cultivate highly remunerative deals. More particularly, however, the developer, with the aid of professional expertise, will have prepared a detailed financial breakdown of the proposals on the basis of which his negotiating position will be well founded. In conspicuous contrast, the local planning authority has invariably had to resort to 'guesstimating' possible inclusions and, in doing so, has often undervalued their potential. Specifically, several features of the developer's approach to negotiation should be borne in mind:

1. The development team will include highly experienced negotiators. Indeed, a fundamental aspect of the many branches of the property industry – investment, development, agency and management – is negotiation, much of that cumulative expertise stemming from close professional liaison. The developer will therefore be very well prepared, having been briefed on what the authority may expect of him in the form of inclusions, what other developers have been prepared to accept and whether any appeal precedents have been set. His financial calculations will reflect those expectations

and he will know with some precision at what point his development appraisal becomes marginal.

2. The developer will always present community benefits as financially critical to the scheme that, as a result, is barely viable, while the local planning authority, in their relative ignorance of the financial implications of the proposal, may be hoodwinked into accepting an increased commercial element in order to help pay for, and thus secure, the implementation of the planning gains, an argument based on the premise that the developer pays for gains by reducing his profit margins.

3. The developer will stress strongly the substantial community benefits that the scheme already generates. These will include features that meet local plan requirements, without which the authority would have been entirely justified in refusing planning permission, intangible qualities that cost little or nothing but are incapable of assessment in financial terms and, in particular, community benefits that amount to development gains too. For example, the inclusion of sports facilities in a commercial project, made available to the local community at certain times, may well improve the value of the commercial content by making, say, offices more attractive to tenants (and their employees) who will be prepared to pay a higher rent. Indeed, despite the impression given by the developer that profits on the scheme will be marginal as a result of the community elements, the costs involved will often be offset, not only as noted above, but also, in areas where planning gain policies have been consistently applied, by the local land market adjusting accordingly.

4. Finally, the developer will always have the option of appealing to the Secretary of State and, ultimately, of withdrawing from negotiations altogether, a threat which often contributes to authorities underplaying their hand and is only mitigated where the site in question is the subject of stiff competition between developers, a situation local planning authorities should always seek to imply.

The combination therefore of a favourable decision-making context and tactical awareness provides the developer with a solid foundation for negotiation. It is, however, his detailed financial appraisal that is the source of the greatest advantage, a knowledge of which would significantly improve the authority's position. Although unfortunately only dealt with nominally in the professional training of planning officers, acquiring the methodological skills to prepare appraisals is not difficult. Indeed, development appraisal packages, modified to meet public sector requirements, are available in computer software form (see Planning Gain Consultants, 1988). Attaching values to the input variables is however critical but, if the local planning authority are in a position to make a realistic assessment of potential profit, based on sound local market intelligence, then not only will opportunities for negotiating community benefits be identified but also a reasonable expectation of the scale of such benefits will emerge.

However, in pursuing negotiated solutions which incorporate community benefits such as industrial and business space developments, three particular issues must be recognized and where possible addressed. They are:

(a) The frequency of suitable opportunities
(b) The achievement of real net benefits to the community, rather than compensatory trade-offs
(c) Securing the implementation of planning gains.

With regard to frequency of suitable opportunities, it is important to stress again that the practice of negotiating community benefits varies enormously, in direct correlation to the level of market demand for particular uses. Thus, for example, in Central London where sites suitable for modern offices are in very short supply but demand for modern space is considerable, local authorities are in a strong bargaining position and, in line with paragraph 4.15 of the Greater London Development Plan (1976), have often specified particular requirements in statutory development plans and, in the light of rapidly changing development circumstances, in supplementary guidance vehicles such as development briefs. As an illustration, 'Plan for Camden' stated that 'Office development will be restrained in the area south of Euston Road . . . Exceptions may be made where substantial advantage can be attained, such as. . . .'

In contrast, however, in those parts of the country where the local economy is particularly depressed, the authority may well regard any development as beneficial in itself and would regard negotiated inclusions as a disincentive to much needed development which they could not afford to lose.

It is of equal importance that the true nature of so-called planning gains are appreciated. In fact, it is relatively rare for a local planning authority to achieve a real net planning gain to the community, at no cost to itself. In most cases, planning gains amount to no more than a trade-off in which the authority allow say, a departure from local plan requirements, or permit a more intense use of a site than would otherwise be acceptable, or incur social costs as a consequence of a development project, such as increased congestion on local roads, in exchange for some other community benefits. These bargains necessitate careful evaluation of the social costs and benefits, not least because of the political repercussions of compromising established planning policies. Such assessments are fraught with difficulties and compare unfavourably with the simple financial appraisal of the developer, whose position is particularly strong if he has previously acquired the site for an undisclosed sum. In-house expertise within the authority is often ill-equipped to provide the necessary market evidence on which a sound negotiating position can be built while private consultants, invariably with vested interests in the development industry, are loathe to be seen to be playing for the other side (in contrast to Planning Gain Consultants).

Whether or not real or nominal community benefits emanate from negotiated solutions, securing the implementation of planning gains is an obvious necessity and raises the thorny issue of planning conditions and planning agreements.

10.6 Planning conditions and planning agreements

Planning conditions

The debate concerning the use of conditions attached to planning permission as a means of securing planning gains is long-standing. On the one hand it can be argued that the achievement of planning gain by definition goes beyond the legally acceptable constraints placed on planning conditions, while others, including the Property Advisory Group have sought to limit planning gains to those that are capable of being the subject of a valid condition. These arguments will doubtless continue to complicate the law pertaining to planning conditions, with its heavy reliance on case law though superficially the situation appears clearcut.

Section 29(i) of the Town and Country Planning Act, 1971, empowers local planning authorities to attach conditions to planning permissions 'as they think fit'. Section 30 goes on to explain that authorities may specifically grant planning consent, subject to conditions:

1. Regulating the development of any land under the control of the applicant, or requiring the carrying out of works on such land, if these matters appear to an authority to be expedient for the purposes of the development
2. For requiring the removal of buildings or works or the discontinuance of any use of land authorized at the end of a specified period
3. For requiring that the building or other operations permitted by the planning permission shall be commenced not later than a specified date.

On the face of it, therefore, powers regarding the imposition of conditions are very wide and are further amplified by the statute which allows conditions affecting other land under the control of the applicant and the granting of planning permission for a specified limited period.

In practice, however, the boundaries of validity of planning conditions have been regularly tested in the courts, where judicial interpretation has tended to place much tighter controls on what is deemed acceptable. Thus, for example, Lord Denning determined in Pyx Granite Co. Ltd v Ministry of Housing and Local Government, 1958 (1 QB 554), that: 'Conditions to be valid, must fairly and reasonably relate to the permitted development. The planning authority are not at liberty to use their powers for an ulterior object, however desirable that object may seem to be in the public interest'.

In summarizing subsequent case law, it has been noted that a valid planning condition must at least;

1. Fulfil a planning purpose (see R. v London Borough of Hillingdon, ex p. Royco Homes, 1974 [2 WLR 805])
2. Be fairly and reasonably related to the development for which planning permission is granted (Pyx Granite case; but see Grampian Regional Council v City of Aberdeen, 1984 [JoL 590] where the planning condition (granted in negative form) was upheld even though it related to danger at a road junction one and a half kilometres distant from the development site)
3. Be clearly reasonable, that is to say, it must not be 'so unreasonable that no reasonable planning authority could have imposed it' (see Mixnam's Properties v Chertsey District Council, 1964 [1 QBD 214] (Heap, 1987).

The increasing use of planning conditions and subsequent challenges in the courts, inevitably prompted a reaction from the Secretary of State, who produced DoE Circular 1/85 on the subject. As well as stressing the useful role that appropriate conditions can play in facilitating development, the circular contained extensive advice, model conditions and six tests, designed to ensure that conditions are 'seen to be fair, reasonable and practicable'. The tests, which are fully explained in the Circular, are:

(a) necessity for the condition;
(b) relevance of the condition to planning;
(c) relevance of the condition to the development to be permitted;
(d) enforceability of the condition;
(e) precision of the condition;
(f) reasonableness of the condition in all other respects

Obviously, securing the details of many planning bargains falls outside the legal limitations applied to planning conditions and therefore necessitates the utilization of a more formal implementation vehicle, the planning agreement.

Planning agreements

Planning agreements represent a form of contract and as such, provide an enforceable means of securing the implementation of community benefits. Intended as a supplement to planning control, planning agreements enable a local planning authority and a developer to formally affirm that matters relating to the use and development of land shall be regulated in a manner which could not be imposed by the authority when exercising its normal development control functions.

An early reference to planning agreements was contained in the Housing, Town Planning, etc. Act, 1909 and enabling powers were subsequently included in section 34 of the Town and Country Planning Act, 1932. The Town and Country Planning Act, 1947 required specific ministerial approval of individual agreements before they could take effect but, with the removal of direct supervision following the 1968 Act, the use of agreements increased significantly.

The principal provisions concerning planning agreements are now found in section 52 of the 1971 Town and Country Planning Act and permit the authority to enter agreements which permanently or temporarily restrict or regulate the development or use of land within the authority's area. Furthermore, they may contain such 'incidental or consequential provisions (including provisions of a financial nature) as appear to the local planning authority to be necessary or expedient for the purposes of the agreement'.

Although section 52, when tested, has proved difficult to construe, as noted by Lawton L.J. in the case of Windsor and Maidenhead Royal Borough Council v Brandrose Investments Ltd, 1983 (1 AII ER 818), the general intent and effect of section 52 agreements have been likened to the powers of an adjacent landowner entitled to enforce a restrictive covenant against his neighbour. Indeed, the restrictive (negative) nature of the provisions of the agreement is vital to the validity of the agreement. [see Abbey Homesteads (Developments) Ltd v Northamptonshire County Council, 1986 (278 EG 1249].

Under separate legislation, however, local planning authorities do have powers to enter agreements containing positive covenants in which, for example, a party may be required to carry out specified works (such as the construction of business workshops). Principally, section 33 of the Local Government (Miscellaneous Provisions) Act 1982 makes suitable provision but similar powers are included in section 38 of the Highways Act, 1980 and many local Acts, as well as the catch-all section 111 of the Local Government Acts, 1972, which enables local planning authorities to do anything 'calculated to facilitate, or is conducive or incidental to' the carrying out of any of their functions. In practice, reference to some or all of these provisions is often included in an individual agreement.

This is in part explained by the lack of clarity in section 52 itself. Although the section refers to the term 'agreement', the effectiveness of such arrangements is based on their contractual nature and, in particular, the remedies available for breach of contract in private law, rather than the statutory Planning Acts, namely, enforcement via an action for damages, specific performance or an injunction, as confirmed in Avon County Council v Millard, 1986 (PL 211). To achieve contractual status requires either consideration (which could arise if, for example, the local authority provided services to the land in question but is not implied by the grant of planning permission) or, by incorporating the provisions in a deed and thus creating a contract under seal. In fact, the legal practicalities involved in development projects and funding agreements are such that planning agreements are almost always made under seal and bind the parties and the landowner's successors in title.

Once established, planning agreements are registered as a Land Charge and exist in perpetuity, unless otherwise stated but, like any contract, can be modified by further agreement. In the case of negative covenants, however, the Lands Tribunal can exercise its discretion under section 84 of the Law of Property Act, 1925 and modify, alter or discharge obsolete or unreasonable conditions.

Planning agreements provide local authorities and the development control mechanism with a significant extension of their administrative discretion, over and above the normal modifications or alterations which the local planning authority may properly seek to planning applications, such as improved access to the scheme. In particular, effective and enforceable contractual powers facilitate the process of ensuring that developers implement otherwise marginally viable or loss-making community benefits. Avoiding the limitations on planning conditions, provisions embodied in planning agreements can be negotiated and confirmed formally, subject to but in advance of the grant of planning permission and, where market conditions permit, have been widely utilized.

The increasing incidence of development deals in growth areas, involving planning gains and, in particular, the use of off-site gains, wholly disassociated with the main scheme under consideration and/or the incorporation of financial payments into the bargain (funds which have been recycled in their entirety, in accordance with capital spending restrictions, unlike other local authority capital receipts), have inevitably attracted considerable professional, media and public attention. Accusations of trading in planning permissions and bribery have arisen, initially promoting the Property Advisory Group report in 1981 and, subsequently, the publication of DoE Circular 22/83 entitled Planning Gain. In seeking to limit the worst excesses of planning gain, the Circular attempted to adapt the planning gains, by focusing on the reasonableness of asking a developer to accept an obligation over and above his development proposals. According to the guidance, this depends substantially upon whether what is required:

1. Is needed to enable the development to go ahead; for example the provision of adequate access, sewerage and sewerage disposal facilities
2. In the case of financial payments, will contribute to meeting the costs of providing such facilities
3. Is otherwise so directly related to the proposed development that the development ought not to be permitted without it; for example, the provision of car parking in or near the site
4. Is designed in the case of mixed development, to secure an acceptable balance of uses.

If the obligation meets one of these tests, then (a) the extent of what is required must fairly and reasonably relate in scale and kind to the proposed development, and (b) what the developer is being asked to provide, or help to finance, must represent a reasonable charge on the developer.

Opportunities for judicial interpretation of the DoE guidance have been infrequent (in itself indicative of the mutual benefits associated with negotiated solutions) but, in the case of Bradford City Metropolitan Council v Secretary of State, 1986 (278 EG 1473), Lloyd L.J. concluded that a planning condition requiring road widening at the developer's expense would not have been lawful if incorporated in a section 52 agreement. This strict interpretation of Circular 22/83 was probably inappropriate but did demonstrate the Courts general attitude to planning gain.

Although many local planning authorities use the tests of reasonableness as a basis for discussion, the guidelines are obviously open to widely differing interpretations and therefore considerable scope for negotiation remains between the parties. Indeed, since 1983, the practice of bargaining has increased, a trend which reflects the potential advantage to developers, who view negotiation as an opportunity to achieve greater returns on their investment at minimal cost and the perceived advantage to local planning authorities, many of whom consider planning gain to be the only financial means available to them to meet local needs and development objectives.

Meanwhile, public scrutiny of bargaining has become more vigilant and, with the advent of Freedom of Information legislation, planning agreements, including financial payments, are open to public inspection and challenge, with third parties seeking to have planning permissions set aside on the grounds that the local planning authority took account of immaterial considerations. The recent circumstances surrounding the case of the Royal Opera House provided a surprising rebuttal to one such challenge, in which the Covent Garden Community Association failed in its High Court action against the decision of Westminster City Council to grant planning permission for 250 000 sq.ft of commercial floor space, on the grounds that it constituted the only way the ROH could fund its own improvements and extensions.

Nevertheless, the continuing high profile of the planning gain debate has prompted some local authorities to be more cautious in their approach and many planning agreements now involve third party agencies, such as housing associations and community trusts, which act as vehicles for channelling community benefits. These arrangements place the authority at arm's length from the proceedings and may foil further Government efforts to apply more sweeping constraints on bargaining practices.

10.7 Conclusion

Since 1947, the role of planning control in implementing development proposals, including industrial and business space schemes, has been largely restricted to responding to planning applications in terms of the Development Plan, after the landowner/developer has identified demand and carried out a development appraisal. Where planning policy and market forces are in harmony, appropriate development has taken place.

Despite efforts to pursue a more positive approach by, for example, preparing development briefs, the development control process is invariably at the beck and call of market forces in implementing specific proposals. The economic recession of the early 1980s and, in particular, its regional dimension, vividly demonstrated the limited abilities of the planning control system to promote industrial activity in depressed areas, while, conversely, more buoyant economic conditions in the late 1980s, especially in the southeast and notably for non-industrial uses, have illustrated the difficulties facing planning control when it attempts to resist market forces even with the support of statutory development plans which may favour industrial activity. The Government has encouraged this trend, in part because reduced or relaxed local control reflects their market-oriented, ideological outlook – the market knows best – while, at the same time, enabling a further centralization of power and, in part through a lack of political will or indeed inclination to any clear image of our environmental future, towards which development should be channelled, preferring instead short-term economic and political expedients.

In response, local planning authorities are beginning to adopt a more flexible entrepreneurial approach, in which negotiated solutions will be countenanced. If the basic principles of the Development Plan remain intact, the heat is taken out of conflicting market expectations and the threat of successful appeals is minimized. Ironically, developers are happy to exploit development opportunities which may arise from uncertainty and emerge through negotiation but, where the stakes are particularly high, for example in the race to succeed in implementing enormous out-of-town retail and associated user projects, the development lobby are advocating more, rather than less, government guidance to local planning authorities and the development control process, in order to minimize development risk.

References

Ambrose, P. (1986) *Whatever Happened to Planning?* Methuen, London.
DoE (1985) Lifting the Burden (Cmnd 9571), HMSO, London.
DoE (1988) *PPG note 5; Simplified Planning Zones*, HMSO, London.
DoE (1988) *Planning Policy Guidance (PPG note 4) Industrial and Commercial Development and Small Firms*, HMSO, London.
DoE Circulars 22/80 Development control – policy and practice.
―――― 22/83 Planning gain
―――― 16/84 Industrial development
―――― 1/85 Use of conditions

—— 2/86 Development by small businesses

—— 13/87 Changes of use of buildings and other land

—— 25/87 Simplified planning zones, HMSO, London.

DoE and Welsh Office (1975) *The Dobry Report: Review of the Development Control System*, HMSO, London.

DTI (1985) *Burdens on Business*, HMSO, London.

Hall, P. (1988) The coming revival of planning. *Town and Country Planning*, February.

Heap, D. (1987) *An Outline of Planning Law*, Sweet and Maxwell, London.

Marsh, C. (1988) *The Use Classes Order*, Pepper, Angliss & Yarwood, London.

Planning Gain Consultants (1988) *Ready Reckoner + Computer Software*, PGC, London.

Property Advisory Group (DoE) (1981) *Planning Gain*, HMSO, London.

Reade, E. (1987) *British Town and Country Planning*, Open University Press, Milton Keynes.

Royal Town Planning Institute (1982) *Policy Statement on Planning Gain*, RTPI, London.

Wates, N. (1976) *The Battle for Tolmers Square*, Routledge and Kegan Paul, London.

11

Implementation by local authorities through land ownership

11.1 Introduction

Chapter 8 examined the growing interest of local authorities in their local economies particularly over the last decade, and how this concern has grown from a desire to ensure the provision of premises to a much wider involvement in all aspects of economic development. Nevertheless the main thrust, at least until very recently, has been in industrial development. The first part of this chapter examines how and why this activity has changed during the post-war era from simple beginnings when land was acquired and parcelled up into fully serviced plots (still a mainstay of many local authorities involvement in industrial development) through to the intense period of direct development of industrial premises in the latter half of the last decade and more recently to an emphasis on a wider strategy and greater involvement of the private sector in a variety of partnership agreements and fiscal incentives from central government. The second part of the chapter examines the main alternatives open to a local authority of implementing development on land within its ownership and their respective advantages and disadvantages. The concluding section examines the process of achieving implementation once a decision has been made on the method to be employed.

11.2 The involvement of local authorities in industrial development

The post-war period until the early 1970s

Many local authorities were involved in industrial development before the last war and several major authorities obtained Private Act powers to develop industrial estates and build factories as a means of lessening the high levels of unemployment that then existed (Barrett and Boddy, 1979). In the immediate post-war period local authorities were again directly involved as a contribution to post-war reconstruction and modernization. The general assumption by the then Labour Government was that both population and economic growth would be low (largely based on pre-war experience), that there was a need to redress the regional imbalance by controlling the location of new industrial employment (the major employment sector) and that most new development would be of an overspill nature to be located mainly in planned new towns surrounding the major conurbations (so avoiding urban sprawl). Much of the overcrowded inner areas of existing cities would be redeveloped by the public sector on new housing estates and, due to the lower densities employed, new towns would need to be built, also by the public sector, to provide homes and employment for the displaced inhabitants (Hall, 1975).

Owing to the assumption of low population and economic growth and the level of public sector intervention, it was anticipated that only a minority of developments would be carried out by the private sector – hence a justification for the comprehensive physical land use development plan system with its strong negative power of control introduced in 1947.

However, this interventionist role for the public sector was short-lived in the sense that relatively successful Keynsian economic policies, combined with the post-war baby bulge, resulted in unprecedented economic and population growth, increased mobility in industry, commerce and population. This caused a continued drift to the southern half of the country and the south-east particularly, the rapid growth of the service sector and most importantly a much greater role for the private sector than expected. The election of a Conservative Government in 1951, heavily committed to private enterprise and the sweeping away of many existing controls, reinforced these trends and resulted in a greatly diminished role for local authorities in the field of industrial development for the next twenty years.

A further reason for the low profile of local authorities during the 1950s and 1960s was the establishment of a regional development policy embracing grants and incentives and the growth of central government agencies such as English Estates (previously English Industrial Estates Corporation) and more recently the Scottish and Welsh Development Agencies. So even in the development areas industrial development was largely undertaken by central, rather than local, government as one of many incentives to firms to locate there. A further factor was that in the southern half of the country it was central government policy, through the use of Industrial Development Certificates, to limit industrial development. Most local authorities, particularly in inner urban areas, viewed industrial buildings with disfavour, especially those located in residential areas. The improvement and increased provision of housing was seen as the primary objective of most authorities, often to the detriment of industrial employment (Heraty, 1979). With relatively low levels of unemployment in most of southern England this was not so extraordinary as it perhaps now appears.

The emphasis on the private sector in the south, and central government agencies/regional policy incentives elsewhere, meant that local authority involvement was generally limited to site assembly and the provision of services and infrastructure so that sites could be made available, usually to industrial firms rather than speculative developers. This activity was naturally more common in the development areas and has been prevalent throughout the post-war period. In addition there were also a few examples of partnership and direct development schemes but these were usually necessitated to house firms displaced either by road widening schemes, or comprehensive development areas of slum clearance, or where they were classified as non-conforming users. The GLC, for example, developed a number of industrial estates in the 1950s for these reasons (Barrett and Boddy op. cit.), although some authorities were less concerned about firms displaced by these schemes and did not ensure suitable premises were made available.

Where local authorities embarked on direct development the projects were usually funded internally or by straightforward borrowing, rather than the more elaborate methods often used in the late 1970s and 1980s. Partnership schemes were also very different from the type of arrangements commonly used more recently. In fact the word 'partnership' is really a misnomer as most of the agreements were little more than leasehold disposal with the ground rent being determined prior to disposal and often fixed for the duration of the ground lease (usually 99 years, although some early examples were shorter than this). So the ground rent was not related to actual, but only expected, costs or rents and rent reviews were not incorporated into leases until the 1960s. Even then they were initially at infrequent intervals of about fourteen years, later shortening to about seven years by the end of the decade in common with occupational leases.

The 1970s

During the 1970s the restrictive attitude towards industry and industrial development in the south slowly began to change at central and local government level with increased concern about rising levels of unemployment and the problems of the inner areas of the larger towns and cities. Initially (in the late 1960s) this concern about inner urban areas focused on social problems but in the 1970s, particularly after the severe recession of 1974/1975, the underlying problem was gradually perceived to be economic.

A number of national initiatives resulted from the Labour Government of 1974-1979 such as the National Enterprise Board, planning agreements, the industrial strategy, the 1975 Community Land Act and the 1978 Inner Urban Areas Act. DoE circular 71/77 specifically required the active involvement of local authorities to facilitate industrial development and together with subsequent legislation generated an increase in local authority involvement in economic development activity. At that time intervention by local authorities was clearly politically acceptable and desirable. As a result local authority policies, particularly in inner urban areas, changed quite dramatically in the latter half of the 1970s and increasingly protected existing firms, helped them to expand, and provided aid to small firms and new firms as they were considered to be important employment generators (following the Bolton Report of 1971). The building of suitable factory units whether directly or with private sector involvement was merely part of a wide ranging interventionist social and economic strategy. Full-time Industrial Development Officers were appointed and co-ordinated at county level. Most of these officers reported to the chief planner (48%) or chief executive (36%) and relatively few (12%) to the chief estates officer or valuer. This indicated 'the degree to which authorities are developing a broader, more positive policy function as opposed to the purely acquisition/letting management functions traditionally carried out by estates departments' (Barrett and Boddy, op. cit.).

The reason behind these strategies was the need to do something positive to combat rising unemployment, but this was not just caused by a declining rate of economic growth in the early and middle years of the decade, but also due in part to previous local authority and central government policies themselves as mentioned above. Comprehensive slum clearance in the 1950s and 1960s (and early 1970s) often meant that factories and workshops were demolished along with unfit housing. Often these industrial premises were in poor condition, but when these areas were rebuilt no factories were put back (due to the dislike of non-conforming users); the owners were either compensated or relocated. Some authorities built industrial units specifically to house these displaced firms but many did not and as a result firms went out of business and jobs were lost. Central government regional policy and the use of Industrial Development Certificates (IDCs) discouraged the growth of firms in the south and encouraged them to relocate in the development areas. Another reason, in London particularly, was asset stripping by large firms. The high price of many industrial sites caused by their potential to be redeveloped for office or residential use provided an incentive to many firms to move, often with government assistance, to cheaper areas.

A wide ranging interventionist strategy was thought essential by many authorities and part of this strategy was the provision of small factories generally of less than 5000 sq.ft in size but sometimes as small as 500 sq.ft. The main reasons behind such active involvement in development were to attract new jobs to an area, to aid new firms to start up or expand, to improve the general environment of run-down industrial areas, to improve an area's 'psychological image' with its industrialists, to achieve knock-on effects to the local area, to ease relocation problems on redevelopment and to gain greater control over tenant selection and hence the type and magnitude of employment (Morley, 1981). But perhaps the overriding reason was because the private and public sectors had neglected the small unit market, because in previous decades there had not been much of a market from tenants (certainly for new units at market rents) and also because of the greater risks involved, the poor security offered by the tenant's covenants, and the management problems involved from a potential high turnover of tenants, etc. In short, speculative small unit developments were unfundable by the financial institutions who, in the aftermath of the property crash of 1974–75, dominated the investment and development market.

Local authorities therefore generally achieved implementation either by direct development or by partnership schemes where they took much of the development risk away from the developers. Often this was by taking a pre-let themselves, which at the same time made the schemes more fundable. Alternatively, but less commonly, they would form a true side-by-side partnership and share the risks and rewards in some predetermined way or guarantee the developer a stated return on costs, leaving themselves with the exposed top slice income. In many such cases the completed and let development would then be sold to an investor.

In more attractive locations, usually green field sites, developers were more prepared to take risks and guarantee the local authority a secure bottom slice ground rent, with equity participation on completion, and frequent ground rent reviews thereafter. With a stronger property market, and high inflation in land prices in the late 1970s, this was a popular option outside inner city areas as it enabled local authorities to achieve some control over development, as well as participate in rising land prices, rather than selling freehold prior to development at the point of greatest risk to the developer (McCarthy, 1980). (See Chapter 14 for examples of such agreements.) But for small unit developments, particularly in inner city locations, this was not an option and frequently local authorities decided to go it alone and become developers.

In 1978 it was estimated that 68% of London boroughs and metropolitan districts and 57% of local authorities generally were building factory units which were below 5000 sq.ft in size and a further 24% were considering doing so (URBED, 1978). This compares with an earlier survey of 1971–72 which showed that only 5% of authorities had new factory units available (Camina, 1974). A number of authorities had very extensive programmes of direct development. For example, in the West Midlands conurbation the County Council and five out of the seven metropolitan borough councils collectively provided over four hundred units amounting to almost 750 000 sq.ft in the late 1970s (Morley, op. cit.). In London there was similar activity with the GLC, most inner London boroughs and many outer boroughs having extensive development programmes. A GLC committee report (GLC, 1979) gave an idea of the magnitude of this provision, stating that in Summer 1978 1700000 sq.ft was available or likely to be available by 1980, in inner London Boroughs (in units under 5000 sq.ft in size). The corresponding figure for outer London was 1300000 sq.ft.

	GLC (sq.ft)	Borough (sq.ft)	Private sector (sq.ft)
Inner London	274761	1056600	390708
Outer London	—	1115146	219999

These schemes were funded by a variety of means. Many later ones were funded by the Inner City Construction Package introduced with the 1978 Inner Urban Areas Act, others by the use of capital receipts, revenue, or covenant schemes but few by borrowing through the LDS allocation (ADC, 1983) (Chapter 12). The LDS allocation had shrunk so dramatically during the decade that there was little surplus to permit industrial development.

Interestingly local authorities undertook direct development, whether they were Conservative or Labour controlled although, because the authorities with the worst problems tended to be Labour controlled, more Labour controlled authorities took such an interventionist stance. Sometimes Conservative controlled authorities could justify such intervention by the financial returns that resulted, securing additional revenue as well as additional employment.

The London Borough of Wandsworth provides an interesting example of an interventionist strategy of the latter half of the 1970s and what happened when political control changed at the end of this period, mirroring the changes that went on nationally at virtually the same time.

The Council's first policy on employment was introduced in a Planning Committee Paper of 1972. Amongst this Paper's recommendations were that local development plans should consider the possibility of attracting new employment in the traditional industrial areas of the Borough, that changes from industrial use should be resisted, that industrial employment should be encouraged, *that industrial estates should be built*, that representations should be made to the government about changing regional policy and that an Industrial Development Officer should be appointed. Nevertheless, by 1976 it was considered that these policies were not positive enough, that housing still received preference over industrial employment where they were in conflict, and similarly town centre redevelopment received preference over industrial use. One central policy conclusion was:

Existing planning powers, concerned with land use, could lead to the preparation of endless maps, diagrams, booklets and reports stating what *ought* to happen – they did not provide the means to *make* it happen. Nowhere was this contradiction clearer than in relation to industry. (L.B. Wandsworth, October 1976).

As a result, a number of firm recommendations were made which formed a much more positive interventionist and wide ranging employment policy. There is insufficient space to list here all the recommendations made by the Policy Review Sub-committee but some of the more important are listed below:

That the Council, in co-operation with other London local authorities would seek to take on a major role in the co-ordination and implementation of economic planning, and in particular should seek a stake in the Planning Agreement System; that the Council should create an Employment Development Office; that the Council should support industrial co-operatives and co-ownership schemes by giving every encouragement to groups seeking to promote such ventures, by direct financial help to embryonic co-operatives, and by inaugurating an Industrial Common Ownership Development Programme under the full-time responsibility of a principal officer in the Employment Development Office; that the Council should be empowered directly to employ staff for industrial production in industrial premises built by or owned by the Council.

It can be seen clearly from the above recommendations that direct development of industrial buildings was rather a small cog in a large wheel although it was part of a coherent policy of 'providing the means to make it (new employment) happen'. It is interesting, though, to realize that a policy of building factories was part of the earlier and weaker employment policy following the 1972 report but that this particular policy took a number of years to get off the ground due to considerable internal argument.

It was not until 1976 that the five-year industrial rolling programme was finally approved by the Council. It comprised nearly 200 000 sq.ft of factory space in units ranging in size from 300 sq.ft to 3500 sq.ft (except for one larger unit) and created nearly 900 jobs.

Wandsworth's programme of direct development was very successful in so far as the units let quickly (with control over the type of tenants occupying them and hence the number and type of jobs provided), firms were prevented from going out of business due to redevelopment and at the same time the Council apparently achieved an acceptable financial return. Nevertheless, according to the 1979 Employment Policy Review Borough Plan topic paper, the main reason for Wandsworth undertaking these developments was because the private sector was not building small factory units on a speculative basis. When the Conservatives gained control of

Wandsworth after the elections in 1978 the policy of direct development was changed.

The new administration was led by young Thatcherite conservatives who believed in minimizing the role of the local authority. Despite this ideology the existing programme of direct development was completed not so much as a means of job generation but because of the financial return achieved. The valuers became more involved and the policy was to obtain the best rent and best references. However, no new schemes were undertaken as the administration's ideology was to reduce intervention and remove the constraints to enable the private market to respond to market demand and a policy of lower rates was implemented to make Wandsworth more attractive to private sector investment. The new administration was encouraged in this view by the upturn in the property market in the late 1970s as shortages of space, due to the low level of development activity following the property crash of 1974–75, caused rents to rise. But, despite the upturn in the market, small industrial units were still not attractive to the private sector. Stimulating development of these units became a priority for the new Conservative national Government which was elected in 1979, as a means of fostering growth of small firms and this became even more necessary with rapidly rising levels of unemployment in the recession of the early 1980s. Their approach to this problem mirrored that of Wandsworth's and as a solution to the problem they turned to tax and grant incentives to the private sector rather then encouraging intervention by local authorities.

The 1980s

As indicated above the 1980s heralded a very different economic and political outlook. The deep recession of the early 1980s caused dramatically increased levels of unemployment, particularly in manufacturing industry, but the political response was very different from that of the preceding decade. To some it seemed that there were similarities with the mid-1970s in so far as the economy was weak, the demand for industrial property was depressed and private sector developers were therefore reluctant to become involved in industrial schemes or inner city renewal. The slow but growing disenchantment with property as an investment, particularly industrial and office property, by the financial institutions only served to reinforce this pessimistic view. Increased, or at least continuing, direct intervention by local authorities therefore seemed a natural result – merely continuing the role started in the mid-70s. However, the Conservative Government considered such direct intervention as politically undesirable. The Government's philosophy was, wherever possible, to reduce the creeping hand of the state, lessen controls and bureaucracy and allow market forces to return, so that the private sector could provide a solution to the problem. It was considered that this would provide a much healthier and sounder basis for economic growth and a prosperous future.

A DoE report of 1982 echoed by the Property Advisory Group (PAG) Report of 1983 clearly stated this new mood:

> 'Local authorities have a range of statutory powers to assemble land, to deal with unsatisfactory conditions, and to bring about desirable renewal and development. These powers can be used on their own account, but often more fruitfully as a means of restoring confidence and of opening the way for private investment which would not otherwise take place . . . local authorities should be seeking to create conditions in which the private sector is able and willing to provide assets and to hold them thereafter.'   (DoE, 1982)

The local authority role was seen as primarily one of land assembly and disposal to the private sector wherever possible – in some respects a return to the situation which prevailed in the 1950s and 1960s. However, this report did acknowledge that in some areas a partnership approach between the private and public sectors was necessary, requiring local authorities to retain an interest in land and buildings, but even then it warned against local authorities underwriting private sector risks, presumably by means of taking leasebacks, as this was seen to be tantamount to additional public sector expenditure which would add to inflationary pressure:

> 'Joint schemes in which the risks of failure are ultimately carried by the public sector partners and in which the returns to private sector investors are in practice secure are unlikely to meet these criteria and stand to be treated as public sector investment projects, which would be counted as such in the context of restrictions of public expenditure.'

Many new initiatives were introduced in the early years of this period to encourage the role of the private sector and lessen or bypass the role of the state. For example Free Ports, Enterprise Zones, Urban Development Corporations, Industrial Building Allowances, Urban Development Grants and the abolition of Industrial Development Certificates, Office Development Permits and the Community Land Act, together with various circulars reducing local authority planning control.

One of the first actions of the new Conservative Government was the abolition of Industrial Development Certificates (IDCs) in 1979. This removed one hurdle for the private sector in the south-east as government approval prior to the submission of planning applications was now no longer required for industrial development. IDCs were originally introduced after the end of World War II as a means of discouraging industrial development in the prosperous south-east and as part of regional policy to encourage development and employment in areas of relatively high unemployment. With high levels of unemployment throughout the country and negative economic growth, the continuance of such a policy was hard to justify particularly to a government intent on reducing bureaucracy.

The second major change was a reduction in planning control. Circular 22/80 and subsequent circulars gave much greater emphasis to the private sector and the likelihood of achieving planning permission. This further reduced public sector control over property development and the working of the market (Chapter 10).

Specific measures to encourage the private sector followed with the introduction in 1980 of 100% Industrial Building Allowances (IBAs) for small factory units up to 2500 sq.ft in size for a three year experimental period, later modified and extended for a further two years. This enabled a developer to offset against tax the full development costs (excluding land) of a scheme. For high tax paying individuals (the marginal rate of tax was 75% until 1984) this was clearly a very significant incentive and resulted in a dramatic change in private sector provision of small industrial units. This important incentive is discussed more fully below. The introduction of 100% IBAs coupled with the much greater controls over local authority expenditure introduced in 1980 (as discussed in Chapter 12) lessened the need, and made it more difficult for many local authorities to intervene in the property market in the way that they had in the latter half of the 1970s.

The introduction of Urban Development Grants in 1982 was a further encouragement to the private sector to undertake development which previously would not have been of interest and was a further way of using public sector money to reduce development risks to such an extent as to make development viable. UDG's and the newer URG's which remove local authority involvement (both replaced by City Grants in 1988), were discussed in detail in Chapter 9.

Finally, mention should be made of disposal notices which were introduced in the 1980 Local Government Planning and Land Act (section 98). The Act gives the Secretary of State for the Environment the power to direct a local authority to dispose of land in their ownership to the private sector in whatever way he sees fit and is yet another measure aimed at reducing local authority control and encouraging the private sector. Directions have been issued to sell land by auction without reserve and sometimes without previously determining the planning position (Searle, 1987). The threat of the use of disposal notices has, not surprisingly, encouraged authorities to act more positively but also, coupled with controls on capital expenditure, to dispose of land to the private sector which they might otherwise have developed themselves or in partnership with the private sector.

As stated earlier, central government envisaged that the role of local authorities should be reduced where possible to doing the minimum necessary to enable the private sector to develop. Even in designated areas within the urban programme – areas with the worst urban problems – central government funding is increasingly difficult for direct development and there is increasing DoE pressure to involve the private sector. In 1982/83

34% of partnership and programme area money spent on economic projects went on the development of industrial units, but this had fallen to 17% by 1984/85 and 15% by 1986/87 (DoE, 1986 and 1988). Over the same period general environmental works, site works, general training schemes, direct support to firms and general advisory services for business all showed an increased percentage share of economic projects. Despite this trend to favour and encourage the private sector through a relaxation of planning controls and the provision of grants and other financial incentives, most local authorities have been unwilling to ignore rising levels of unemployment and have tried to find ways of encouraging new investment by supplying land and buildings. Across the country as a whole the provision of industrial and commercial sites is still the most significant economic development activity undertaken by district councils (ADC, 1987(a)). Although there was a slight reduction in this activity between 1982 and 1984, it has subsequently stabilized at the 1984 level where 75% of district councils provide industrial sites. Similarly, despite pressures imposed by central government, the provision of premises by district councils has increased from 51% in 1982 to 71% in 1986, predominately in units of under 2000 sq.ft (ADC, op. cit.). This activity is likely to continue to increase following the ending of IBAs (outside of enterprise zones) in 1985.

The trend of the 1980s has been to try and reduce direct intervention by local authorities and the Conservative Government has tried to achieve this policy objective by the measures outlined above. At the same time Labour Party policy, both nationally and locally, has moved in the opposite direction in the sense that direct intervention in firms to overcome poor investment, poor management, race and sexual discrimination is now considered more important than simply intervening to build more factories (Fothergill, Monk and Perry, 1987). The use of Enterprise Boards by a limited number of radical authorities and particularly the Metropolitan authorities, before they were abolished in 1985, is a clear example of this approach. For example the London Borough of Haringey's Enterprise Board Corporate Plan 1985/86 states among its objectives:

Allocate resources and develop programmes to ensure that the employment needs of black and minority ethnic groups, women and people with disabilities are met;
Assist the emergence of new forms of industrial ownership and control.

However, even an Enterprise Board set up by a Labour-controlled local authority still follows the prevailing ethos from central government: 'the shortage of financial resources points to the importance of leverage, i.e. using HEB money to secure the maximum possible contribution from the private sector.'

Of all the measures introduced by the Conservative Government, the one having the greatest impact in terms of encouraging the private sector in the provision of industrial units was undoubtedly 100% IBAs. Because of its importance as a means of using public resources to encourage a private sector, rather than a local authority, response, it is worthwhile examining what was involved and how effective it was. Despite increased action by local authorities in the latter half of the 1970s, and particularly following the Inner Urban Areas Act of 1978, a 1980 report (Coopers & Lybrand Associates/Drivers Jonas, 1980) identified a shortage of small industrial premises throughout the country. Initially, when introduced in the Finance Act of 1980 for a limited three year period, only industrial development comprising permanent units of up to 2500 sq.ft qualified for the 100% allowance (larger units qualified for an initial allowance of 75% plus an annual writing down allowance of 4% but this was merely an extension of the system of allowances introduced twelve years before). It was small units of this size that local authorities had concentrated on building in the latter half of the 1970s as the example of Wandsworth illustrated. In 1982 the definition of qualifying uses widened to include some service sector activities (storage for industrial products, etc.) reflecting the severity of the economic recession, its effect on manufacturing industry and the growth in the service sector, but also the successful lobbying from agents of developers who had built units which were proving hard to let as insufficient background market research had been undertaken into their location. In March 1983, when the original qualifying period for IBAs ended, the scheme was extended for a further two years, but only for smaller units of up to 1250 sq.ft in size. Refurbishment schemes were now also included where the *average* size of units was below 1250 sq.ft even though some individual units were larger.

The effect of the 100% IBAs was to dramatically increase private sector provision of small units as shown in Figure 11.1.

But most of the units provided by the private sector were close to the maximum size permitted and shortages of very small start-up units persisted (DOI 1982) which is why the scheme was extended with a lower floor space limit. It was the very small and least viable units which local authorities were still forced to develop themselves (average size of unit in 1981 for example was 1000 sq.ft). Even the units built by the private sector often required head leasing by local authorities to reduce risks and help secure funding. It is interesting that the developers of these units were not the traditional property companies or financial institutions but often individual high taxpayers, or consortia of high taxpayers, who were developing on their accountants advice more as a way of reducing their annual tax bill rather than as a long term property investment. This helps to explain why adequate market research was not always undertaken. Whilst the success of local authority schemes in the late 1970s illustrated the viability of these developments and helped to encourage a private sector response once tax inducements were available, when the 100% IBA system ended in 1985 (except in Enterprise Zones) yields rose, viability was questioned and

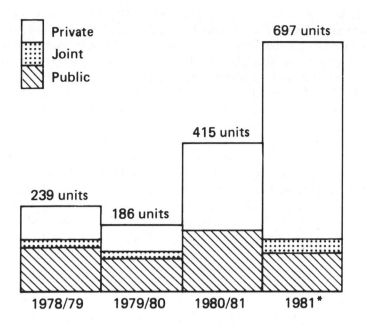

Figure 11.1 Starts of small industrial units by financial years. Note: the data are for all new schemes and conversions in the 22 local authority areas surveyed.
Source: Department of Industry 1982.
*(April to November 1981) × 12/8.

private sector development declined. As a recent ADC survey shows there is now a danger of an acute shortage (Association of District Councils, 1987(b)).

Figure 11.1 shows that local authority involvement in the provision of small units was much reduced compared to the pre-100%-IBA era. Nevertheless, local authorities continued to develop very small units themselves or in partnership with the private sector on a lease and leaseback basis. For example in 1983/84 it was estimated that local authorities directly developed 230000 sq.m of industrial units, which was nearly 15% of the total of new units provided by the public and private sector combined (Fothergill, Monk and Perry op. cit.). Where authorities developed themselves schemes were often financed by deferred purchase agreements with finance houses. At the time (pre-July 1986) these agreements were outside the capital expenditure control system in so far as they effectively spread the capital cost of a scheme over many years so that annual capital expenditure was not significantly increased. Other financing methods involving local authorities leasing sites to finance houses, developing a scheme as an agent and then leasing back the completed development did count as prescribed expenditure but the capitalized value of rents received often offset the capitalized value of the rent paid (before the percentage of capital receipts that could be used to boost prescribed expediture in the year of receipt was reduced to 50% and then to only 30%). Furthermore, the finance house could benefit from the 100% IBAs as it held a leasehold interest which in turn meant that the local authority benefited indirectly through reduced interest charges (for a fuller discussion of controls on local authority expenditure and borrowing see Chapter 12).

An alternative approach, increasingly of interest to local authorities, has been the use of 'arms length' agencies called economic development companies. As explained in Chapter 12 as these companies are legally separate from their parent local authorities they are outside the Government's expenditure controls and so can raise finance from the private sector, transfer expenditure from one year to another and are not constrained by the regulations governing capital receipts (but see Chapter 12 for discussion of DoE proposals to alter these regulations). Additionally they can be more flexible and rapid in decision making (Planning Exchange, 1987).

The 100% IBA scheme achieved the Government's objective of increasing the supply of small industrial units as a means of encouraging the growth of small firms. It also demonstrated the success of using public money to subsidize, and therefore stimulate, the private development sector rather than local authorities. Nevertheless, it was an indiscriminate and open-ended subsidy and did not encourage a detailed assessment of demand by developers. Much of the assistance was therefore probably poorly targeted, leading to over-development and high vacancy rates in some areas and insufficient supply in other areas where it was perhaps more needed but less viable (Adams, 1987). Perhaps as a result of this experience the City Grant requires (as did its predecessor, the Urban Development Grant) a very careful assessment of the need for each scheme by the DoE. There are obvious benefits of this approach but also problems. Firstly, it relies on a willing developer; it involves considerable time and bureaucracy and there is no way of enforcing the take-up of grants once they have been agreed. It also tends to encourage marginal schemes in the more attractive and least difficult areas to develop (Hart, 1984).

Where local authorities own little developable land, City Grants can be a useful means to encourage implementation, but where they own land and viability is questionable some form of partnership arrangement, usually a lease and leaseback, is the favoured solution by many authorities. For very small units, or where demand is weaker, or the location is less attractive to the private sector, direct development is still pursued by authorities but generally only as a last resort. At least the Government now appear to recognize that this is necessary 'for the foreseeable future because of the continuing reluctance of the private sector to invest in this kind of activity. It appears that private sector institutions are no longer prepared to make this sort of investment, even in the least depressed part of the assisted areas. . . .' (DoE 1986(b)). But for many authorities direct development is only possible where capital assets can be sold to finance a rolling programme, as controls on capital expenditure make this very difficult otherwise. As a result, some authorities are unable to meet demand for premises and have been forced into a slowing down of activity (Rowan-Robinson and Lloyd, 1987). Where demand is stronger land assembly and subsequent disposal continues as a mainstay of local authority involvement in economic development. Most local authorities now provide a comprehensive package to promote economic development which will involve the private sector wherever possible and not just for industrial development.

'The proposed strategy for Council action will provide a framework within which to exploit Wandsworth's economic development potential to the full by promoting all types of industrial and commercial property development, enabling the most effective use of available private sector investment and guiding the allocation of central government and Council resources.' (L.B. of Wandsworth, 1986).

The changes to the Use Classes Order and the consequential potential loss of industrial premises to business use, and the loss of industrial land to residential and retail warehouse use, coupled with the economic recovery and increased output by manufacturing industry, point to possible shortages of industrial premises in the near future. Once again this may well require greater direct intervention by the public sector to provide space and/or further tax and grant incentives to encourage the private sector to do so. For small industrial units these shortages are already evident in many areas. The ADC Survey referred to above (1987(b)) showed that 73% of respondent local authorities reported a shortage of units of 500 sq.ft or below and 63% between 500 and 1000 sq.ft; 44% said that the demise of the IBA had had a direct effect on the private provision of small premises under 1250 sq.ft.

Central Government, responding to this problem and intent on reducing the role of local authorities, has encouraged English Estates to embark on a programme of small unit industrial developments in inner city areas throughout the whole country but Lord Young has emphasized the need for rents to be set at economic levels so that a reasonable return on capital is achieved; the ultimate aim being that rents must increase so that eventually they will reach a high enough level to encourage the market to operate freely without the need for public sector intervention. This seems unlikely to occur in the areas that are most in need and flies in the face of the pump priming role that such schemes have performed, as explained in Chapter 7.

At the macro level the Government's new found enthusiasm, since the 1987 general election, for the inner cities and urban renewal has resulted in a rash of mini-UDC's (Chapter 9) in preference to giving local authorities greater powers and finance through the Urban Programme for example.

(The Chancellor of the Exchequer's 1988 Autumn statement shows a £17 million cut in the Urban Programme but increased funds allocated to UDCs.) Nevertheless some entrepreneurial authorities such as Birmingham have initiated large scale partnership arrangements with the private sector through the formation of joint companies.

Birmingham Heartlands is an Urban Development Agency, or joint company, between the City Council and Bryant, Douglas, Galliford, Tarmac and Wimpey. The chairman is an ex-Conservative MP and the chief executive is a former chairman of Tarmac. The company's aim is to regenerate a very large area of nearly 2 350 acres north east of the city centre. The development strategy involves a programme of new infrastructure, land assembly using the City Councils CPO powers where necessary to overcome fragmented land ownership, the redevelopment of about a quarter of the area for business, industrial and residential use over a 10-year period, the marketing of development opportunities and negotiation of City Grants, an SPZ (Special Planning Zone), job retraining and community improvements. In many ways it is like a UDC but with direct local authority involvement and a wider brief.

Another form of partnership between the public and private sector is the Phoenix initiative which was launched in 1986. This is a non-profit making, non-party political, private sector led organization aiming to promote public and private enterprise in urban renewal and regeneration through an organizational and financial partnership using public statutory powers. An overriding aim of the initiative is to work on a grand strategy avoiding isolated schemes and creating a virtuous spiral. To be successful therefore the private sector must be involved with its finance and expertise. An example of one of these initiatives is at Teeside where Phoenix are trying to set up a partnership between the Development Corporation and up to five local authorities. Other initiatives have been in Manchester, Salford, Bristol and the Wirrall.

Both these examples of partnership illustrate current trends in urban renewal but go beyond the scope of this chapter because although local authorities are involved, implementation does not necessarily depend on local authority land ownership. In some areas where large areas of derelict land are in local authority ownership more traditional partnership agreements are still possible however. An example of this is in Swansea where the local authority own a 170 acre site and are negotiating a partnership agreement for an industrial/residential scheme with a consortium called British Urban Development (BUD) comprising eleven major UK construction companies.

The joint company approach which has been growing in popularity may be shortlived if the government's proposals contained in the 1988 Consultation Paper are enacted (Chapter 12). Despite the advantages of this approach and its enthusiastic support by the private sector, the Government want to severely limit local authority involvement in such companies as it regards them as purely a device to circumvent controls on local authority capital expenditure.

The theme of the first part of this chapter has been to trace the growing involvement of local authorities in implementation through land ownership and how in recent years various attempts have been made by central government to limit this involvement and encourage the private sector to act alone. A variety of methods have been adopted to carry out this policy but probably the most effective has been the controls on capital expenditure. This is such a complex and important subject that much of Chapter 12 is devoted to explaining how the system works, how it has changed, how local authorities have attempted to get round the restrictions and how these loopholes have gradually been eliminated. But before this can be undertaken it is necessary to examine the methods and procedures that local authorities have adopted in implementing development where land is in their ownership.

11.3 Alternative methods of implementation

Where a local authority owns land various alternatives are available to achieve implementation. Not all these alternatives will necessarily be possible on all sites all of the time; depending on the circumstances some may well be impossible but usually there will be a degree of choice with advantages and disadvantages attached to each alternative. The alternatives range at one extreme from outright freehold or long leasehold disposal to the private sector in return for a capital payment, to direct development with no private sector involvement at the other extreme. In between there are a variety of possible different partnership arrangements which in themselves cover a spectrum of local authority risk taking from four-slice arrangements, traditionally used in many town centre and prime office schemes, where most of the risk is taken by the private sector; to two-slice lease and leaseback arrangements where most, if not all, risks are taken by the local authority. There are yet other arrangements which do not neatly fall into any of these three categories where, for example, the local authority has no land ownership but yet either guarantees a rental income to the developer or goes further and takes a head lease (i.e. pre-let) itself.

Before examining these alternatives in more detail and their advantages and disadvantages it is worth mentioning why private sector developers are prepared to be involved in many of these arrangements. What potential benefits do they receive? First, they can utilize local authority powers of compulsory purchase which can often, but not always, make land acquisition quicker and cheaper or merely enable it to occur at all. Occasionally without CPO powers or the threat of their use, land assembly might be impossible if a key landowner refused to sell. Whilst CPO powers are perhaps most useful in large town centre schemes where there is often a myriad of land ownerships they can also be helpful in other areas, for example on inner city sites where it may be desirable to marry a number of separate ownerships to obtain a larger development site with better access and more flexibility of development. Secondly, the developer will be working with the local authority rather than against it, which should lessen conflict, enable the developer to benefit from a wealth of background information and research, ensure local authority backing for the scheme and guarantee planning permission. Thirdly, as the local authority have a financial stake in the development, through land ownership, this arguably encourages them to be more commercially minded and support a more market orientated commercially viable development. Fourthly, where local authorities ground lease a site, rather than dispose of it outright for a capital sum, the developer will have a lower capital outlay and will not have to incur expenditure up front. This will mean a reduced borrowing requirement and possibly reduced risk.

Outright disposal

Despite the advantages of partnership arrangements to the private sector, in many situations outright disposal is favoured as it increases private sector control over development and enables funding to be secured more easily as freehold or long leasehold interests (subject to a peppercorn rent) are favoured by banks, insurance companies and pension funds. For large high risk schemes, e.g. Canary Wharf, partnership arrangements where profits are shared might deter the private sector from being involved and might even jeopardize the whole development (National Audit Office, 1988). Possible future problems concerning refurbishment, adverse gearing between rack rental income and ground rent, and voids are lessened where freehold ownership is retained by the private sector.

For example, where the private sector have a long lease and rent payable is a significant percentage of full rental value and is reviewable on an upwards only basis (typical of partnership agreements on central area developments in the 1970s and early 1980s) problems will occur if rack rental income falls, due either to falling rental values or voids (when leases end or exceptionally where tenants go bankrupt). The ground rent must still be paid as a first charge out of income, and will not be reduced if it is reviewable on an upwards only basis. These potential problems are compounded during refurbishment, where there may be voids whilst work is in progress and where considerable sums of money may need to be spent by the private sector leaseholders to update their investment. The capital expenditure should increase rental values and therefore ground rent payments to the freeholder who may not have had to contribute anything.

The present Government, in marked contrast to the previous Labour

Government, also favour the approach of outright disposal and have increasingly used the provision contained in Section 98 of the 1980 Local Government, Planning and Land Act to issue disposal notices to local authorities which it considers are retaining land unnecessarily and where the land could be developed by the private sector. Because of the possible benefit to the private sector of outright disposal it is often argued that developers will pay more – i.e. that the capital sum will be greater than the capitalized value of the ground rent that would be payable under a partnership agreement (Hedley, 1985). Furthermore some valuers argue that a higher sum would be obtained for a freehold compared to even a very long leasehold of say 125 or 150 years, as the freeholder will have even greater control than a long leaseholder. The RICS Planning and Development Division guidance notes on Development Briefs reinforce this view: 'Where a freehold sale of land is involved the value of the land will be maximized if any restrictions on the development of the land can be minimized.' (RICS, 1985).

Some landowners (British Rail for example) who particularly want to retain an interest in a scheme have 'assessed what they would have received in ground rent and taken this directly in the form of buildings on their own land rather than as shared income. They have benefited from taking, as it were, 100% of a part of an estate rather than, say, 15% of the whole'. (Hedley op. cit.). The freehold part retained then gives the owner total control to choose tenants according to commercial or other criteria of its own choosing. However, whilst some writers advocate this approach others advocate against outright disposal even from a financial viewpoint (McCarthy, 1980). It is argued that to sell the freehold of a site prior to development commencing is to sell it at the point of greatest risk. A higher price would be obtained by entering into a partnership agreement with a developer and then selling when the scheme has been completed, thereby sharing in the success and profitability of the development. This is particularly relevant where land values are rising rapidly in the upturn of the typical development cycle. Nevertheless, in today's financial and political climate such a choice may be academic as disposal notices (or the threat of them) and the need for capital receipts (even though capital expenditure can only be increased by a proportion of the receipt) will dictate the decision. But where there is a real choice the main reason why many local authorities would not favour an outright sale is the lack of control over subsequent development and the risk that once land has been disposed, development might not occur in the form desired with the new owner attempting a more favourable planning permission, perhaps on appeal. Furthermore, for certain types of development such as nursery units and workspace developments or development on inner city sites, outright disposal may not even be an option as the risks of development would be considered too great by the private sector and some form of risk sharing partnership agreement would be essential. In the more disadvantaged areas direct development by the local authority may be the only solution.

Partnership agreements

As mentioned above, there are many different types of partnership agreements although in many of these it is debatable just how much of a true partnership has actually been created. Most of these agreements involve a division of responsibilities at different stages of the development process and risk is rarely shared equally. Arguably a true partnership would only occur with a joint company approach. So the term 'partnership' is used to embrace any agreement between a freehold land owner (freeholder) and a developer (leaseholder) where the risks and returns of development are shared in some way, however unequally. A general rule is that risk and return go hand-in-hand but this is an over-simplification.

What are the advantages of a partnership approach to the local authority? First, partnership agreements enable local authorities to act positively and take steps to initiate planning policies and implement development, rather than preparing plans and waiting for developers to respond. In many cases partnership agreements not only enable control over the timing of development but they enable initiation of schemes which otherwise would not have got off the ground due to the uncertainties and risks involved, or the type of development and its location. In some, admittedly rare, cases however local authority land ownership and the possibility of a partnership agreement could discourage the private sector from site assembly and initiating development for fear that after much time and effort piecing together part of a potential development site, the local authority might then compulsorily acquire their interest and may subsequently not even appoint them as developer where a competitive tender is used. However it is more likely that where a developer owns a large part of a potential development site, it could be in a very strong bargaining position to resist the compulsory purchase of its landholdings and to force the authority not only to nominate it as developer but possibly on advantageous terms as there would then be no competitive tender (this is discussed in more detail later in the chapter). In this situation the CPO powers (or the threat of their use) can be an advantage to a developer in terms of the acquisition of the remaining sites not already in its ownership.

As well as enabling local authorities to initiate development, land ownership gives them great powers of control both over the form of development, including its letting, and over any future reletting, refurbishment or redevelopment during the period of the lease. Greater control is achieved because a landlord can exert greater control than can be achieved through planning legislation. Restrictions on development control caused by government circulars and a developer's right of appeal, the limited use of conditions on the grant of planning approval and problems over the inclusion of planning gain can, in theory, be overcome. Many local authorities cite lack of control as the major drawback of freehold disposal. They fear that once land has been disposed, different uses or mix of uses, or even size of units within the same use, will be applied for. Even if planning permission is initially refused it may well be achieved on appeal. The new Use Classes Order will undoubtedly reinforce these worries because of the greater freedom given to landlords within the new business use class (Class B1) and the ability to transfer from Class B2 to B1 (given by the new General Development Order).

Whilst partnership arrangements should ensure greater control over development, they nevertheless still retain the benefits of involving private sector expertise and experience in the development process. By nominating a development partner, after a selective competitive process, the authority in theory gains choice as well as expertise, ensuring the best overall scheme in terms of layout and design as well as in financial terms. This brings possible advantages over direct development (where greater control compared to outright disposal is also achieved) because the private sector has greater experience in development and has access to finance which is uncontrolled by central government (Chapter 12).

The question of financial return is a complex one and is inextricably tied up with risk, the type of development and its location; different types of partnership agreements must be considered separately before any conclusions can be made. For well located institutionally acceptable office and industrial schemes three- or four-slice partnership agreements, or increasingly side-by-side agreements subject to a minimum ground rent, will usually be possible where the freeholder receives a secure ground rental income which is a first charge out of income generated by the development. This rent will be regularly reviewed, sometimes in an upwards only direction, and there will be participation clauses to enable the freeholder to share in any excess profit generated by the development (see Chapter 14). In this situation the freeholder will participate in future rental growth, which in part it will help create through the general prosperity of the area and the success of the scheme where large scale re-development is involved. Ground rental income should therefore increase until ultimately, when the lease ends, the whole development will revert to the freeholder and full rental value will be received directly from the occupational tenants with no payments to any intermediate leaseholders.

This type of agreement, if the development brief and lease are properly drawn up, should enable the freeholder to achieve a secure ground rental income without undertaking the risks of development, which are borne by the private sector almost entirely. Where local authorities are the freeholder they can therefore be seen to control development more effectively for the benefit of the community and participate in its rewards without taking many risks. However, whether the financial return will be as high as an outright freehold disposal is questionable for the reasons mentioned earlier. During a development boom where land values are rising rapidly then a local authority may ultimately achieve a better financial return by retaining a regularly reviewable ground lease with an equity participation clause – but where land values are not rising significantly the greater benefits to a developer/investor of freehold ownership and greater control over the development and subsequent investment will probably result in a higher price from outright disposal.

However, for industrial development particularly, unless the scheme and its location are good, this type of partnership agreement will not attract the private sector, mainly because they would shoulder most of the risk. The main worry is tenant demand and tenant covenants. To exaggerate, with a freehold if there are 100% voids there will be zero income, with a leasehold there will be negative income and the higher the ground rent the greater that negative income will be. Even in a prime scheme it is therefore unusual to find developers willing to pay a ground rent which is more than about 10% of full rental value (unless it is a side-by-side agreement). Any excess would normally be capitalized and paid as a premium.

One possible solution is not to have a horizontally divided three- or four-slice arrangement but a vertical side-by-side arrangement with risks shared more equally and actual income, whether it rises or falls, split between the parties in some predetermined way. But this too has its problems although it is increasing in popularity for prime schemes. As there is no clear division of authority, particularly over letting, problems could arise over approval of tenants and whether covenant or the level of rent is more important.

The most common form of partnership agreement in high risk areas of high risk types of development, such as nursery unit industrial schemes or workspace developments, is some form of horizontal two-slice agreement where the local authority takes the more exposed top slice, partially guaranteeing the developer a reasonable or sufficient return on costs. In its crudest form the developer may be allowed to receive a stated return on costs with the residue going to the local authority when the scheme is typically sold on to an investor or the occupiers. But in many inner city areas finding an investor on these terms may be difficult! Most commonly, therefore, local authorities retain the freehold and take a leaseback from the developer of the completed scheme and are responsible for sub-letting themselves. The developer receives a pre-let to one tenant of undoubted covenant, which reduces development risks and makes funding easier. The leaseback rent is usually geared to full rental *value* and reviewed in an upward only direction, thus ensuring a reasonable return if costs are properly controlled. The local authority takes most of the risk because if rental *income* falls due to voids (almost inevitable in small unit schemes) its profit rent will suffer (see numerical examples in Chapter 14). But on the other hand it achieves total control over letting. In theory it can choose which tenants to accept, the terms of the lease and the level of rent. Although local authorities are duty bound by Section 123 of the 1972 Local Government Act to achieve the best possible price, or rent for leases of more than 7 years, this regulation is qualified by the phrase 'having regard to all the circumstances.' Some authorities therefore 'constrain a lease in such a way that the level of market rent is affected and the policy objective legally and properly achieved' (Searle, op. cit.). This can be achieved in many ways. Commonly many authorities will let on short leases (maybe 5 years or less) with break clauses; some incorporate 'good employer' clauses in leases/licences or attempt to use lettings as a means of support for priority industrial sectors (Cook, 1987).

Lease and leaseback agreements are not without problems however. As mentioned in Chapter 12, under the 1980 Local Government, Planning and Land Act controls over capital expenditure, both income received and rent paid would be capitalized and counted as a notional capital receipt and capital expenditure respectively. Prior to Circular 5/87 even if the former was larger than the latter and a profit was made, only part of the capital receipt (originally 100% but later 50% and more recently 30%) could be counted in one financial year, so that effectively an increase in capital expenditure resulted with all the problems that this entailed. This situation is now slightly improved following Circular 5/87 and 100% of the notional capital receipt can be used to offset the notional capital expenditure pro-viding they occur in the same financial year (but see Chapter 12 for a fuller discussion of this point and additional problems caused by ministerial state-ments in early 1988).

A further possible problem surrounds the agreement between developer and local authority of what the full rental value actually is, as this deter-mines the developers income and therefore development return. Where a local authority exercises its control over letting in the way described above, actual income received from occupational tenants may differ from full market rental value and disputes will arise. Another problem could occur where the developer miscalculates the cost of development and finds himself in the position where costs are rising but potential income, which is determined by full rental value, is not. This means that profit margins will be eroded providing a powerful motive to reduce costs by reducing stan-dards, to the detriment of the local authority.

Because of these potential problems, but more likely because of the pri-vate sector's unwillingness to develop due to the location, type or size of scheme or state of the property market, direct development by local autho-rities or public sector agencies could be the only solution. It should be emphasized, however, that in the present climate of reductions and greater controls over local authority capital expenditure and grant/tax incentives to the private sector, this is very much an action of last resort. The London Borough of Islington, for example, which had an active programme of direct development in the late 1970s continuing on a reduced scale in the recession years of the early 1980s, reached a policy decision in 1984 that no further floor space schemes would be developed by the Council. The only rare exceptions to this policy being:

1. In the case of Council owned sites or buildings, where both the following factors apply:
    (a) There is clear evidence that the private sector is not prepared to undertake a similar development;
    (b) A development can be demonstrated to be cost effective in terms of job creation or job retention.
2. In the case of sites or buildings in private sector ownership, acquisition/ development by the Council will be an extremely rare occurrence and will only be considered where there is very clear evidence of the cost effectiveness of doing so in terms of job creation or job retention. (L.B. Islington, 1984).

Where direct development is necessary or desirable it will enable local authorities to achieve greater, i.e. total, control over development both in its initiation and over the development process itself. They should receive a greater financial return as they shoulder all the development risks and they should be less vulnerable than in a partnership scheme where the developer may be able to exploit his monopoly position once he has been nominated following the normal procedure of short listing and competitive tender. Norwich City Council cited this concern as one of the main reasons that they undertook development in the 1970s as they were worried that, once selected, a developer might subsequently present very good reason why the original scheme could not proceed and negotiate a revised more commercial scheme with planning gain elements reduced (Minns and Thornley, 1978). It is even possible that an unscrupulous developer might deliberately overbid initially on the cynical assumption that it could utilize its monopoly position once nomination was assured (Linacre, 1980).

There are of course potential problems with direct development just as there are potential benefits. In their 1980 report Coopers and Lybrand considered that over-zealous development in some areas had crowded out the private sector. Increased supply had reduced rents (to the benefit of tenants) and made schemes unviable to the private sector in locations where otherwise they would have developed (Coopers and Lybrand Associates/ Drivers Jonas, 1980). Ambler and Kennett were concerned as to whether local authorities have a formal procedure for appraising projects and consider that sometimes the only constraint has been whether the budget can take it. Where a commercial return is supposedly achieved they cast doubt over whether all costs have been included in the financial appraisal

(Ambler and Kennett, 1985). Undoubtedly there are potential problems with direct development and probably the main ones are the risks involved and the lack of expertise and proper management structure to ensure a successful development.

> 'It is not practical for a local authority to carry out this role of entrepreneur. A committee should not be asked to take a commercial risk of this kind, answerable as it is to full Council and to the ratepayers as a whole. Shareholders expect their money to be at risk, albeit remotely; ratepayers do not.' (Powell, 1974)

The irony here, of course, is that in many partnership agreements local authorities have to take a leaseback of the completed development and therefore are taking most of the risk. Direct development is one stage beyond this but by using design and build arrangements, or employing project managers, many of the potential problems of the development process can be lessened. A revised committee structure with delegated authority to the relevant chairman and nominated chief officers can also reduce the internal management problems. What is important is that there should be a clearly defined programme of implementation with an internal management structure to go with it, rather than a few ad hoc schemes with no proper control. This was one of the problems encountered by the London Borough of Islington in the early 1980s, contributing to their decision not to develop any further schemes except in exceptional circumstances (L.B.Islington, October 1983).

In conclusion, it is worth emphasizing that local authorities can take a longer-term view of viability compared particularly to trader developers in the private sector. Immediate profitability is less important if the scheme is not being sold. Subsequent rental growth may convert a marginal scheme into an attractive financial investment. It is ironical that local authorities are often considered not to be equipped to undertake a risk-taking role, which is essentially a private sector activity, yet it is the very situations where the private sector will not develop due to the greater risk involved, that local authorities have been forced to undertake development as an aid to employment regeneration and urban renewal.

11.4 Methods and procedure of implementation

Partnership arrangements

Each scheme will be organized slightly differently depending on the size and type of development, but the following programme gives a general idea of the main stages that are usually undertaken although the precise order may vary:

overall policy decision to undertake development

appoint consultants where in-house expertise is insufficient
decide overall size and composition of scheme
design scheme for initial appraisal. This may be done in-house or by
    using consultants
initial appraisal
amend scheme
preparation of development brief
selection of development partner
preparation of detailed scheme
detailed planning permission
discussions with statutory undertakers
prepare CPO if necessary
negotiation to obtain land by agreement
public inquiry and confirmation of CPO where relevant
serve notices
relocation
vacant possession
demolition
design agreement or final details
building agreement
construction and cost meetings
letting meetings and letting
participation/leaseback rent calculations

Most of the stages in the above programme are self-explanatory but two
require further discussion – the development brief and the selection of a
development partner. The objectives of a development brief can be summarized as follows (RICS, 1985):

promote interest in the development of a site
provide positive planning guidance over land
give the landowner reasonable control over the form and content of the
scheme
provide a clear basis for the design of a scheme on which developers can
work
set out the main financial terms required by the landowner
inform the developer of what plans and written material should be
submitted
provide a common basis for comparing developer's proposals
minimize the scope for renegotiation after tender date.

As mentioned earlier in this chapter, one of the main benefits of partnership arrangements to a local authority compared to freehold disposal is the
greater control over the form and content of development. One of the objectives of the development brief is to spell out to prospective developers the
type of uses, their floor space, site coverage and layout principles including

aspects such as access points, landscaping, circulation and loading space requirements, where the scheme is an industrial development. This is important not just to ensure that the development proposals conform to the local authority's wishes but also to provide a common basis for comparing developers' proposals. However, whilst this is desirable in most cases it may not always be the best solution particularly on large sites where a variety of alternative uses may be acceptable to the local authority or where they do not want to constrain a developer's entrepreneurial flair and ideas. The problem with this more flexible approach is trying to select the best scheme where so many different criteria are involved.

Another alternative at the opposite extreme is for the local authority to design the scheme itself and ask the development partner merely to submit a financial bid. This approach makes comparison relatively easy but is rarely used because developers dislike it as there is no scope for any layout or design flair to enable one developer to produce a better scheme and therefore outbid competitors.

One final alternative, again rarely used, is for the local authority to fix the financial terms – ground rent, premium etc. and let the developers compete on layout/design grounds. This approach is disliked by the local authorities as they cannot be certain that they are achieving the best possible financial terms. If the terms are pitched too high, potential bidders will be put off; if the terms are pitched too low, either excess profits will accrue to the selected developer or else the scheme will be built to an unnecessarily high specification. The financial terms contained in development briefs, and relevant to partnership schemes, are discussed and illustrated in Chapter 14.

The process of selecting a development partner is crucial to the success of a partnership scheme. The criteria that should be employed in selection are:

the financial offer (ground rent and/or premium) and equity share
the sustainability of the financial offer (i.e. is it an overbid, are costs and rents reasonable, etc.)
the layout/design/appearance of the proposal from an architectural, planning and estate management point of view
analysis of developer's previous schemes
the track record of the developer, its financial standing and other commitments
the financial backing of the developer for the scheme
the developer's team of advisers and consultants
the personalities of the development team and particularly of the key personnel of the development company.

Clearly, therefore, even with a detailed development brief, the financial offer is only one aspect to be considered. The above criteria are more applicable to a competitive tender although they are also relevant when negotiating with a single developer. In most cases it is desirable from the local authority's point of view to have a restricted competitive tender to ensure the best possible scheme and the best financial offer, thereby complying with S.123 of the 1972 Local Government Act. However, there are occasions where it may be appropriate for the local authority to negotiate with one developer, where that company has a substantial ownership in the proposed site or where there is insufficient interest from other developers to make a competition worthwhile or possibly where one developer has particular expertise and is well known to the authority. Normally a competitive tender will be held; the site would be advertized to maximize interest in the scheme and a shortlist drawn up from those developers expressing firm interest. Depending on the degree of interest the short listing will be in one or two stages. If there is considerable interest a preliminary short list of about ten companies would be drawn up by the officers often with consultancy advice. These companies and their advisers would be interviewed usually by elected members, to enable a second short list of three to five companies to be produced. A limit on the number of developers preparing schemes and financial offers is necessary, unless the scheme is simple and small in size, because otherwise the time, effort and expense would deter developers becoming involved if their odds of selection were less than about 20–25%.

Direct development by local authorities

The case has often been made by the private sector particularly that direct development by local authorities is inappropriate and undesirable for many reasons mainly due to the problems over finance, risk, lack of expertise and experience, and the infrequent committee cycle which prohibits quick decision-making. There is undoubtedly truth in these observations, but they fail to take account of specific changes in internal organization that could and indeed have occurred in many authorities that have undertaken development themselves and the different approaches to implementation which are possible. For example, whilst a local authority could undertake the development entirely by itself using in-house staff and an outside contractor this is not usual particularly for larger schemes. Sometimes schemes are designed and costed in-house but often outside architect and quantity surveyors with greater experience are used, just as a development company uses consultants in its development team. It is also common, particularly for industrial schemes, to put the whole scheme out to tender on a design and build basis. The most appropriate scheme can then be selected by the authority and then built and managed by the contractor on a fixed price basis leaving the local authority to look after letting.

The Pilcher Report (DoE, 1975) expressed the view that some large urban authorities have planning and valuation staff who are sufficiently experienced in managing large urban estates and successfully promoting commercial development. These authorities also have senior officers well able to advise members on decisions affecting such development, but according to the report the majority do not, in which case the use of consultants was recommended. This is usual practice in any case where development of any size is to be implemented by using a partnership agreement with a developer.

The reorganization of local government in the early 1970s (earlier in London) paved the way for the introduction of an internal corporate structure based essentially on private sector experience and so it could be argued that although local authorities are much larger than private sector development companies their corporate structure provides them with a small inter-disciplinary development team of chief officers.

'The awareness and sense of the importance of corporate identity as well as departmental responsibility reduces and, in many cases, removes the need to expound the advantages of a development team to handle local authority development work.' (Cadman and Austin-Crowe, 1983)

Where a development team is necessary, its composition must depend on the size and importance of the development programme with subsidiary teams often being established to deal with individual projects. The main team should comprise the Chief Executive, the Planning Officer, the Engineer responsible for highways and transport, the Finance Officer, the Estates Surveyor/Valuer, with other officers (and consultants) for advice where necessary (Cadman and Austin-Crowe, op. cit., p. 240). The team should be kept fairly small mirroring private sector organization; even so conflicts may still occur, particularly between the Finance Officer or Valuer and the Planner. The development team must have a strong leader 'who has a broad understanding of the different professional contributions which must be blended together' (DoE Pilcher Report, op. cit.) and with delegated authority to make quick decisions, where necessary. There should also be a project manager, who could be the same person as the development team leader, on small schemes particularly. In some cases the project manager would be a paid consultant or developer.

Where there is no strong leader, problems such as inconsistent and slow decision-making on day-to-day issues, failure to resolve inter-departmental disputes and lack of forceful project management and co-ordination may well arise. It was these reasons (and others) that forced the London Borough of Islington, for example, to terminate its programme of direct development in 1984 (L.B. Islington, 1984). As a result of these factors development costs were unnecessarily high and schemes suffered from time slippage.

Most successful direct-development schemes have project managers, most have a working party of chief officers and most report to a specially constituted sub-committee (often of the Policy and Resources Committee) which could either meet at very short notice as and when required or give delegated powers to the chairman who maintains regular contact with chief officers. Most authorities, therefore, utilize a streamlined organization to speed up decision-making with some loss of democratic control by elected members (Morley, op. cit.). The Pilcher report was of the opinion that strategic decision-making, general policy formulation and decisions on financial terms must occur in the full Council or main Committee.

> 'but the discussions and negotiations leading to these decisions, the detailed points which must be determined at the intermediate stages, the planning of the scheme, the monitoring of local authority interests during construction are matters which can and often should be delegated to sub-committees, chairmen and officers in a systematic way, according to the level of importance of each issue.' (DoE, Pilcher Report op. cit.)

Unnecessary delay is avoided and important benefits will result, more than outweighing the loss of member involvement.

References

Adams, D. (1987) The nature of demand for small premises. *Estates Gazette*, 1 August.

Ambler and Kennett (1985) *The Small Workshops Scheme*, Department of Trade and Industry Report, HMSO, London.

Association of District Councils (1983) *Economic Development by District Councils. Paper One. Financing Economic Development and Aid to Industry*, Association of District Councils, London.

Association of District Councils (1987a) *Economic Development by District Councils. Paper Six (revised). Economic Development Initiatives and Innovations*. Association of District Councils, December, London.

Association of District Councils (1987b) *A Blueprint for Urban Areas?* Association of District Councils, London.

Barrett, S. and Boddy, M. (1979) *Local Government and the Industrial Development Process*, School of Advanced Urban Studies, Bristol University.

Barrett, S. and Boddy, M. (1980) *Local Authority/Private Sector Industrial Development Partnership*, School for Advanced Urban Studies, Bristol University.

Boddy, M. (1983) Changing public–private sector relationships in the development process. In *Urban Economic Development, New Roles and Relationships* (eds. K. Young and C. Mason) MacMillan, London.

Cadman, D. and Austin-Crowe, L. (1983) *Property Development*, 2nd edn, E. & F.N. Spon, London, p. 238.

Camina, M.M. (1974) Local authorities and the attraction of industry. *Progress in Planning* **3**(2).

Cook, G. (1987) Industrial property letting: traditional v. radical approaches. *Local Economy*, **2**(1), May.

Coopers and Lybrand Associates and Drivers Jonas (1980) *Provision of Small*

Industrial Premises. Department of Industry Report, HMSO, London.

DoE (1975) *First Report of the Advisory Group on Commercial Property Development*, (Pilcher Report), HMSO, London.

DoE (1982) *Bringing in Business*, HMSO, London.

DoE (1986a) *Assessment of the Employment Effects of Economic Development Projects Funded Under the Urban Programme*, Inner Cities Research Programme, DoE, London.

DoE (1986b) *UK Regional Development Programme 1986/90 as Quoted in a Blueprint for Urban Areas*, Association of District Councils 1987.

DoE (1988) *The Urban Programme 1986–87*, DoE, London.

Department of Industry (1982) *Small Workshops Scheme. Survey of the Effect of the 100% Industrial Building Allowance*, DoI, London.

Fothergill, S., Monk, S. and Perry, M. (1987) *Property and Industrial Development*, Hutchinson, London.

GLC (1979) *Industry and Employment Committee Report 121.*

Hall, P. (1975) *Urban and Regional Planning*, Pelican, London.

Hart, D. (1984) *Attracting Private Investment to the Inner City. The Hackney Demonstration Project*, Joint Centre for Land Development Studies, Reading University.

Hedley, A. (1985) *Finance for Development* in *Managing the Local Economy* (eds P. Williams and B. Bourdillon), Geo Books, Norwich.

Heraty, M. (1979) *The Role of the Local Authority and the Changing Fortunes of Inner City Industry*, School of Planning Publications, Polytechnic of Central London.

L.B. Islington (1983) *Review of Industrial Floorspace Policy and Programme Development*, Management Sub-committee, 31 October, London Borough of Islington.

L.B. Islington (1984) *Industrial Floorspace Policy and Programme Development*, Management Sub-Committee, London Borough of Islington, 13 March.

L.B. Wandsworth (1976) *Prosperity or Slump? The Future of Wandsworth's Economy*, London Borough of Wandsworth, October.

L.B. Wandsworth (1986) *Policy and Finance Committee*, Paper no. 3633 on Industrial and Commercial Development Strategy. LondonBorough of Wandsworth, April.

Linacre, V. (1980) Town development – '80's in the shade. *Estates Gazette*, 1 March.

Local Authority Associations (1988) *Stimulating Local Enterprise – The Local Authority Role*, Local Authority Associations, London.

McCarthy, P. (1980) *How to Succeed in Partnership by Trying Very Hard*. 1980 Seminar Paper published by The School for Advanced Urban Studies, Bristol University, Working Paper 18: Local Authority/Private Sector Industrial Development Partnerships.

Minns, R. and Thornley, J. (1978) *State Shareholding*, MacMillan, London.

Morley, S. (1981) *Positive Planning and Direct Development by Local Authorities,*

School of Planning Publications, Polytechnic of Central London.

National Audit Office (1988) Department of the Environment: Urban Development Corporations. Report by the Controller and Auditor General, HMSO, London.

Planning Exchange (1987) *Economic Development Companies*, Local Economic Development Information Service, The Planning Exchange, Glasgow.

Powell, G. (1974) Local authority participation in the profits of local property developments. *Chartered Surveyor Urban Quarterly*, **1**(3), April.

Property Advisory Group (1983) *The Climate for Public and Private Partnerships in Property Development*, HMSO, London.

Rowan-Robinson, J. and Lloyd, M. (1978) *Local Authority Economic Development in Scotland*. Planning Exchange Occasional Paper, 32.

RICS (1985) *Guidance Notes on Development Briefs*. Planning and Development Division RICS, London.

Searle, B.S. (1987) *Valuation and Estate Management – Valuing your Assets*. Paper given at S.T.S. conference on the Effective Management of Land and Property Resources, 26 June, Royal Institution of Chartered Surveyors.

Society of Local Authority Chief Executives/RIBA/CIPFA/RICS (1986) *Our heritage – Property Management in a Local Authority*, Joint Report, London.

URBED (1978) *Local Authorities and Industrial Development*, Urban and Economic Development Group, London.

12

Local authority financial and legislative powers

12.1 Introduction

Local authorities have considerable power to undertake many types of economic development initiatives. Notwithstanding the availability and applicability of these powers, the ability to use them is now severely constrained by restrictions imposed by central government on raising and spending money.

A variety of legislative powers have been in existence for many years but recently they have been widened and clarified. Controls on local authority borrowing and expenditure, however, have been changed and tightened over successive years, particularly over the last decade. This chapter examines the constraints on local authority capital expenditure, how these controls have changed and the effect these changes have had on local authorities' ability to undertake economic development initiatives involving land and property. The chapter concludes by looking at the legislative powers available to local authorities.

12.2 Capital expenditure and finance

Local authority expenditure is classified by central government into two areas – current and capital spending, the former dealing with day-to-day expenditure, the latter with creating fixed tangible assets which will last and therefore benefit the community for a long time. Capital expenditure is now a relatively small part of total local authority spending (about eleven per cent) but in the 1960s and early 1970s it was a much greater proportion. It has declined to about half in real terms of what it was then. Capital spending is partly, but no longer entirely, financed from loans. Other sources are capital receipts, rates, grants, leasing and capital funds built up over many

Table 12.1 Financing of local authority capital expenditure, England and Wales

	1981/2 (%)	1985/86 (%)
Borrowing	70	57
(including advances from capital funds)		
Capital receipts	7	25
Government grants	12	7
Revenue contributions	9	7
(including transfers from special funds)		
Leasing	2	5

Source: DOE Consultation Paper July 1988.

years in some cases (see Table 12.1). When it is financed by borrowing the debt is repaid over a number of years, varied according to the assumed life of the asset. The capital debt is therefore converted into an annual cost which, together with revenue spending, makes up the total of current expenditure. This is financed from the rates (poll tax), government grants and other income – primarily fees, charges and rents. In recent years, not only has capital expediture, declined but so too has the proportion of revenue spending provided by central government through grants (the Rate Support Grant in particular). In the mid-1970s over 65% of local authority current expenditure was provided by central government grants. Now this is down to 46% (1986–87). Out of this diminishing percentage specific grants have increased rapidly in recent years as central government attempts not only to reduce local government expenditure but to exert greater control over certain parts of it.

Expenditure on land assembly, infrastructure works and developing land is clearly capital spending, and the ways in which money can be obtained and the controls exercised by central government over how much can be raised or spent require examination and, in particular, how these controls have changed.

12.3 Central government control prior to 1981

Before 1970 (i.e. before Circular 2/70) authorities had to receive central government permission (loan sanction) before they could borrow money to finance any item of capital expenditure. This cumbersome time-consuming procedure left local authorities with little flexibility, but it did mean that once loan sanction was given it guaranteed that money could be borrowed over a number of years; there was no danger that the allocation for subsequent years would be cut unexpectedly, a situation which local authorities have been faced with more recently.

Circular 2/70 and subsequent circulars during the 1970s fundamentally altered central government control by dividing local authority capital borrowing into two parts – key sector and locally determined schemes (LDS). The former covered borrowing to finance major items of expenditure of national importance such as housing, education, principal roads, water supply and sewerage. Loan sanction was required for each scheme or programme of schemes. The acquisition of land for key sector schemes was classified as a subsidiary sector, and was not subject to loan sanction control once key sector approval had been given. Schemes of more local importance (known as Locally Determined Schemes or LDS), including general land purchases and development expenditure, received a block loan sanction – the allocation being on a county basis (except in London where the GLC's borrowing was controlled by an Act of Parliament) and apportioned by consultation with local authorities in the area. The LDS allocation therefore gave local authorities some flexibility as they could spend this borrowing allocation however they chose. In the early 1970s many schemes of direct development by local authorities were financed out of the LDS allocation (in some cases, such as the shopping centres at Watford, Banbury and Swindon, by using a special part of the LDS quota called the Large Projects Pool, but this was ended by the government in 1974).

After the 1975 Community Land Act, spending on land acquisition for non-key sector activities was transferred from the LDS allocation into a separate key sector account. In theory local authorities had slightly more money allocated for land acquisition but, importantly, less flexibility over how they could spend it. The abolition of the Community Land Act, however, did not lead to a corresponding increase in LDS allocations, and by 1979/80 LDS was only 50% of its value in 1973/4 at constant prices.

Towards the end of the 1970s an additional allocation of money was made to relevant authorities under the 1978 Inner Urban Areas Act. This was called the construction package and was 75% funded by central government. This was the forerunner of what is now known as the Inner Area Programme referred to later. Many local authorities used this to finance direct development of industrial units in the late 1970s.

Local authorities' flexibility and control over capital expenditure was effectively gradually lessened during the 1970s as the LDS allocations were cut back significantly. Many were forced to consider raising money by means outside central government control and this was particularly so for large items of capital expenditure, such as commercial and industrial development.

Methods of raising capital from internal sources free of central government control prior to 1981

For relatively small items of capital expenditure finance from revenue was a possibility, particularly if the authority owned a large stock of revenue producing land and buildings. However, there were obviously limits to the amount that could be raised from revenue, particularly if it came directly from the rates over a short period of time, as it would impose a severe burden on present ratepayers when the capital asset would benefit future as well as present ratepayers. At that time Section 137 of the 1972 Local Government Act (which has subsequently been amended) limited the amount of revenue that could be used this way in any year to the product of a rate of 2p in the pound, but gave authorities a wide-ranging power on what it could be spent on. Sub-section 1 states: 'a local authority may . . . incur expenditure which in their opinion is in the interest of their area or any part of it or all or some of its inhabitants.' However, this section cannot be used to incur expenditure where an authority is authorized or required to do so by any other Act, i.e. a specific function of the authority such as housing.

Many authorities have of course used this power to promote economic development in a wide-ranging way, and this ability to raise money is, at the time of writing, still available, although for most District Councils the amount of money that can be raised is obviously limited. The Government's response to the Widdicombe Committee of Inquiry's Report however proposes major changes to this power (see later in this chapter).

A possible way round the financial limitation of only being able to spend up to the product of a rate of 2p in the pound in any financial year on the loan charges arising from capital expenditure was, and still is (although other restrictions are now relevant), for an authority to set up a Capital Fund under Section 28 of the Local Government (Miscellaneous Provisions) Act 1976, which states that an authority can 'establish such funds as the authority consider appropriate for the purpose of meeting any expenditure of the authority in connection with their functions; and make into a fund established under this paragraph such payments as the authority think fit.'

Capital receipts from the sale of assets, with the exception of receipts from the sale of housing, could be and still are a way of supplementing local authority spending on land acquisition and development and are obviously a way of avoiding borrowing. Capital receipts could also be paid into a fund, as above, to give greater flexibility in the timing of their use. This is still the position today, although local authorities' spending limits can only be increased by 30% of the value of any capital receipts received by selling assets (but see Circular 5/87 referred to later). The sale of existing assets has proved over many years a popular way for local authorities to buy or develop land and buildings.

Other sources of money of a slightly different nature and which did not count against LDS allocation were central government grants or borrowing from the authority's own superannuation funds (applicable to the largest top tier local authorities). Until 1983 there was a limit which prevented more than 25% of a superannuation fund being invested in property, and in any case a commercial return would have to be obtained, otherwise the fund and the future pensions of the authority's employees would suffer. However, if development was expected to be profitable in the long term then this could be an appropriate investment for the superannuation fund, which could invest money in return for an initially low rental return but with the prospect of future rental income growth (just as pension funds generally do in purchasing property investments).

Government and EEC grants such as from the European Regional Development Fund could be used where applicable and did not count against LDS allocations.

Raising capital from external sources free of central government control prior to 1981

Use of internal sources of revenue or capital was an alternative way for local authorities to raise money without having to borrow externally, which would have counted against their LDS allocation. However, there were also a few methods of raising money externally which were not classified as borrowing money, and hence came outside central government control, although in many ways they might appear to have been little different from borrowing. One of these methods, involving transfer of legal interests and titles to land (and therefore not classified as borrowing) was lease and leaseback.

The lease (or sale) and leaseback transaction was a common method of financing commercial developments in the private sector in the 1960s and 1970s, and a number of local authority town centre direct developments were also financed in this way. This method, extensively used in the 1970s by local authorities for industrial and retail development schemes, had two slightly different forms. The first was relevant when local authorities took on the role of developer either of town centre schemes (relatively rare) or industrial schemes. It involved the authority selling a financial institution a long leasehold interest – usually 125 years – for a sum of money equivalent to the cost of development, and the institution granting the authority an underlease (i.e. a leaseback to the authority) for the same period (less a few days) at a rent which reflected an acceptable rate of return on capital invested. The rent or 'interest' charged therefore equated to normal property investment criteria and was reviewed at regular intervals. For a good quality property investment the interest rate was lower than borrowing on the money market, but in return the institution would share in future rental growth and could insist on some control over the scheme's design and letting. A variation of this, used particularly in industrial development, involved a third party as developer with the local authority leasing back the completed development and subletting to occupying tenants itself.

A second method used occasionally in the late 1970s (but more common in the 1980s) to bypass central government control over borrowing was the deferred purchase or covenant scheme. A finance house would agree to build an industrial development on land owned by the authority but without taking a lease. In return the authority agreed to repay capital and interest over a medium-term period (usually 5–10 years). Capital was never therefore borrowed and the finance house had no security other than the good name of the local authority.

12.4 Central government control post-1981

The 1980 Local Government, Planning and Land Act introduced significant changes to central government control over local authority expenditure. From 1 April 1981, control by way of loan sanctions, i.e. control over borrowing, was supplemented by control over spending by way of cash limits. Each local authority's centrally imposed cash limit gives an entitlement of loan sanction for borrowing up to this limit. Some of the methods mentioned previously which enabled local authorities effectively to increase expenditure by bypassing controls over borrowing are now no longer possible. The Government's objective was clearly to restrict monetary growth and public borrowing as a proportion of GDP to help limit, or even eliminate, inflation and to limit public expenditure to provide scope for reducing taxation. This, it was hoped, would improve motivation and efficiency with a consequential increase in employment opportunities. Under this system local authorities have, in theory, increased freedom to decide their own priorities for capital expenditure within overall limits imposed by central government. In practice this has meant more freedom to spend less.

Different levels of local government control different services. Depending on whether the local authority is a County Council, District Council or Metropolitan District Council it has to supply the relevant central government departments with programmes of expenditure for the following year for which of the following five blocks it has responsibility: housing, education, personal social services, transport and 'other services' (in essence similar to the various key sectors and LDS under the previous system). Partnership and programme authorities designated under the 1978 Inner Urban Areas Act have a sixth block – Inner Area Programme. Central government assesses these programmes and then allocates an overall cash limit to each local authority for the relevant blocks under its control. (Most District Councils receive two annual block allocations for housing and 'other services', County Councils four blocks and Metropolitan District Councils all five blocks provided they are an education authority.) These block allocations may be aggregated and treated as one overall cash limit; hence the apparent freedom of authorities to decide their own priorities for capital expenditure, but only within an overall limit set by the Government. However if by so doing some of the blocks are underspent (and others by definition overspent) then the Government may reduce the allocation to the underspent block(s) the following year. So it is arguable just how much freedom authorities have in deciding their own priorities for capital expenditure. There is a tolerance limit which enables the overall cash limit to be under or overspent by up to 10% and the following year's total adjusted accordingly.

Capital expenditure permitted under the 1980 Act is limited to the allocation under the blocks mentioned above plus the 10% tolerance, plus capital receipts and profits from trading undertakings. Capital expenditure is defined in the Act to include the acquisition of land and buildings, the reclamation, improvement or laying-out of land, the construction, improvement or replacement of buildings and the giving of grants or advances of a capital nature. Such expenditure is referred to as Prescribed Expenditure. Clearly then capital spending by a local authority on any industrial or commercial development with which it is directly involved will be Prescribed Expenditure. These stringent controls have therefore tended to encourage involvement of the private sector as well as indirect involvement by local authorities through land disposal and partnership arrangements, rather than direct development as explained in more detail later. However, the situation that local authorities find themselves in is not as straightforward as it may appear. Some activities such as acquiring a lease of private sector development will count as notional capital expenditure even though only a rent is paid. Also, just as there were ways to avoid government controls prior to the 1980 Act, so there were under the new system and a whole industry of creative accountancy developed to help authorities carry out what they see as necessary development in their areas. A close examination of the post-1980 situation is therefore required.

Prescribed expenditure financed by borrowing

First, before examining Prescribed Expenditure (which does not involve borrowing) and expenditure which is not classified as Prescribed Expenditure (and which also does not involve borrowing) it would be useful to examine block 5 and, where relevant, block 6 allocations (i.e. other services and Inner Area Programmes) and capital receipts and profits from trading undertakings because clearly if an authority has sufficient resources under either or all these three headings then it will have considerable freedom to intervene in the local economy.

Block 5 allocations are limited in amount and there are many competing demands from services other than industrial or commercial development. For many authorities extensive development capital programmes will be ruled out by the extent of their capital allocations (ADC 1983), just as they were from LDS allocation in the late 1970s.

Fifty-seven local authorities (thirteen of which are in London) are able to submit Inner Area Programmes (IAP) under the revised Urban Programme. This is a special allocation of resources to local authorities (equivalent to a sixth block) set aside from the normal resources available in the main expenditure programmes to support a range of projects aimed at economic regeneration, improving the environment and dealing with social problems in urban areas. The relevant authorities receive grant at 75% from central government for approved projects, but 75% of any surpluses will be claimed by the DoE. Just how beneficial this extra block allocation is remains debatable. First, many inner areas have lost much more in Rate Support Grant than they have gained in Urban Programme Funds (LGIU May 1987) and so it could be argued that the sixth block has partially or wholly been created out of, rather than being additional to, the other five. Secondly, each project within the programme requires approval and as DoE Circular 17/83 states the public sector should not undertake any projects of commercial development which are capable of implementation by the private sector and in the case of intervention of any kind, its form should be the minimum necessary – and lowest cost – to make the project happen (e.g. as in UDGs and now City Grants).

The traditional Urban Programme where large numbers of authorities outside the main areas of urban deprivation could bid for a 75% central government grant to support relatively small specific projects of urban regeneration was abolished in 1986. Although the projects were small (£160 000 maximum for example in 1983) the 25% local authority contribution received borrowing approval outside the capital spending control system.

Capital receipts

Capital receipts (as Table 12.1 indicates) have been a considerable source of capital to finance economic development as indeed was the case in the 1970s but, of course, this depends very much on the land holdings of particular authorities. Although originally, under the 1980 Act, 100% of capital receipts could be used to supplement capital allocations, this was progressively reduced to 50% in 1983 and only 30% (Wales 50%) in 1985 (20% for housing in England but 15% in Wales). This meant that only 30% of the capital raised by selling an asset could be used to supplement capital allocations in that same financial year and 30% of the remainder (i.e. 30% of 70% which is 21% of the original amount) in the next financial year, etc. This is known as the cascade effect. Circular 5/87 amended these regulations in certain circumstances where notional capital receipts have been received (see below and also the proposals covering capital receipts in the DoE consultative paper of July 1988).

Another major impact of the 1980 Act concerning capital receipts is the concept of notional capital receipts. Leasehold disposal of land (e.g. where a ground rent is received by a local authority landlord in a traditional partnership arrangement) is treated as if the freehold capital value was received if the lease is for more than 20 years. Where the lease is 20 years or less this does not count as a capital receipt unless Prescribed Expenditure is incurred on the acquisition of the land or in connection with buildings on the land. But if such expenditure has been incurred the capital receipt is only scored to the extend of the expenditure incurred (Circular 5/87). Leasehold disposal on a ground rent basis can therefore be a useful way to supplement capital allocations, although prior to 1987 the 30% cascade rule was particularly unhelpful in lease/leaseback situations where a local authority took a lease and sublet to occupying tenants as mentioned later. It should also be remembered that although capital allocations are supplemented by the amount of notional capital receipts, no actual capital has been received and so the way the expenditure is financed will be important as no right to borrow is given.

Finally, the 1980 Act permits local authorities to supplement their allocations from profits on trading undertakings received in the same financial year, profits being calculated on a current cost accounting basis. Some local authorities, such as Leicester DC have a corporate estate which is classified as a trading undertaking and therefore provides regular income which can be ploughed back into land assembly or actual development.

Prescribed expenditure without borrowing

Expenditure of a capital nature on economic development is clearly Prescribed Expenditure under the 1980 Act, as mentioned above. The problem faced by most local authorities is how it is to be financed. If there is some scope in the block 5 allocation (other services) then clearly this will give an entitlement of loan sanction to borrow (although it must not be forgotten that the annual loan repayments still have to be paid out of revenue which may also be subject to Government constraint). However, for most local authorities this is a non-starter as there are too many other competing demands on this limited allocation. Similarly profits from trading undertakings will not be relevant to most authorities as even if they have trading undertakings it is only the annual profits which can be used and in many cases they will be of limited use in paying for capital expenditure. Capital receipts are often therefore the only possibility but only if there is a political will to sell and the authority has a substantial number of assets which it can realise. If it does not, there is a problem, but one which can often be solved if the scheme is profitable – i.e. the income received offsets the costs incurred. As industrial and commercial direct development by local authorities involves the receiving of income this can be capitalized and counted as a notional capital receipt and so be used to provide the Prescribed Expenditure allocation for new projects. The capital required could then be obtained by one of the means described in the following paragraphs. The 30% cascade rule now means that a revolving fund would probably need to be established although Circular 5/87 will affect this. The problem with this approach is that some local authority developments by their very nature of creating employment in locations unappealing to the private sector are often unprofitable.

1. SPECIAL FUNDS AND REVENUE CONTRIBUTIONS

Special or capital funds were referred to earlier as a way of avoiding controls on borrowing. This still applies, although the 30% cascade rule will apply for capital receipts that have more recently been paid into funds. The use of revenue to finance capital expenditure was also referred to earlier and for small items of expenditure this may be a possibility for many local authorities. S.137 of the 1972 Local Government Act is widely and increasingly used by local authorities for economic development with some authorities fully committed to the product of a 2p rate. In 1975–76 only about 35% of all councils appear to have made use of S.137 but by 1984–85 the proportion had risen to 75% and economic development and employment promotion measures were by far the largest single use of S.137 powers (Widdicombe, 1986). Nevertheless, a recent survey showed that, amongst District Councils, the average take up is still relatively low although increasing (ADC, 1987). Section 44 of the Local Government (Miscellaneous Provisions) Act 1982 specifically extended the powers to give financial assistance (by lending, guarantee or making grants) to persons carrying on commercial or industrial undertakings, but the Government's response to the Report of the Widdicombe Committee of Inquiry proposed major changes to local authority powers (see below).

Where local authorities are directly involved in development and the capital sums involved rule out the above methods of financing, lease/ leasebacks (until 9 March, 1988) and deferred purchase/covenant schemes (until July 1986) probably were the most commonly used methods. Two forms of lease/leaseback were mentioned earlier as this method was in use for many years as a way of avoiding borrowing to finance capital expenditure. The 1980 Act as amended by the Local Government Finance Act 1982 affects these variations both of which involve the local authority taking a lease and subletting to occupational tenants.

2. LEASEBACKS

Section 80 of the 1980 Act provides that where a local authority acquires land for a period exceeding one year other than by acquiring the freehold, then the authority will be regarded as having paid the freehold or market value of the land. This means that acquiring land by leasing will count against capital expenditure block allocations, just as if the land had been purchased outright. (Prior to March 1988 the exception to this rule was where the lease acquired was for not more than 20 years except where a permanent building was to be erected at the authorities request.)

So, if an authority who owns land which it wishes to develop itself sells a long lease to a financial institution for a premium, it acquires a capital receipt for it, 30% of which can be used to supplement its annual block allocation in that financial year. If the authority then leases it back, as a means of 'borrowing' capital, the annual rents count as normal revenue expenditure, but the value of the freehold out of which the lease was granted (i.e. equivalent to the total capital sum borrowed) counts as notional capital expenditure against the block allocation. If subleasing to occupational tenants occurs in the same financial year as the leaseback transaction then the notional freehold value of the site can be counted as a notional capital receipt, which can be used to offset the notional capitalized value of the headlease. Prior to 1987 only 30% of this notional capital receipt could be counted to offset expenditure incurred but Circular 5/87 introduced 'back to back' schemes enabling, in effect, the capital expenditure to be matched by the capital receipt providing that the capital receipt could not be generated without the expenditure being incurred and expenditure and receipt occur in the same financial year. (Such offsetting is not as of right and requires Ministerial consent for extra allocation.)

Because of the reduction in pension fund and insurance company involvement in the property market, particularly industrial property, these lease/leaseback arrangements are usually agreed with a financial house and the rent payable is in effect equivalent to a mortgage repayment with a variable interest rate based on six months' LIBOR (London Bank Offered Rate). In such arrangements the aim is to secure to the financier the remaining capital allowances for industrial buildings (this is discussed in some detail in Chapter 11) which means that the lease will usually be at least 30 years and the occupying tenants will have to be industrial users who qualify for the allowances. The benefit of these allowances (4% per annum writing down allowances over 25 years) is rather less now than it was between 1980 and 1985 when 100% IBAs were available for small industrial units. The leaseback arrangement enabled local authorities to benefit indirectly from these allowances by effectively paying interest (rent) at much lower rates than if conventional borrowing had been arranged (which of course was difficult due to the controls mentioned). One problem with this approach is the penalty involved for early redemption which makes it difficult to offer tenants an option to purchase.

A number of variations on the lease/leaseback arrangement are available. It is usual in the version described above for the local authority to act as agent for the finance house in the construction of the units. Commonly the authority will then appoint a builder on a design and build competitive tender basis. Alternatively a developer may be involved on a risk-taking basis and the authority will lease the units (taking a pre-let) from the developer and sublet as before. The developer in this case arranges the finance but the capital expenditure implications for the local authority are similar as if the local authority itself had undertaken the development. The difference between the two methods will relate to the rent that they have to pay. In the finance lease, repayments to the finance house are related to LIBOR and although they may fluctuate they are unlikely to change significantly. Where the lease involves a developer, rent payable by the local authority is likely to be a percentage of full rental value subject to regular review. Where investment yields are below LIBOR, initial repayments will be lower in this approach but subsequent rental growth may make future payments higher. These aspects are considered in greater depth in Chapter 14. Whichever variation is used it is important to avoid the sub-leases being granted in a different financial year from the lease/leaseback agreement. If this happens the notional capital receipt will not be available to offset the notional capital expenditure, and when the notional capital receipt is received only 30% of it can be used in that year to increase the prescribed expenditure limit.

3. HEADLEASING

Where local authorities do not own land, headleasing agreements similar in many ways to those described above, can still be useful as the local authority covenant will usually assist the developer in obtaining finance and risks. If they take a lease for more than three years (prior to March 1988 it was 20 years) this involves notional capital expenditure which, since Circular 5/87, can now be offset by the notional receipt from subletting (assuming it occurs in the same financial year).

However, before the March 1988 clampdown, if local authorities took a lease of the completed buildings for less than 20 years, then this would not count as Prescribed Expenditure provided no contract to build existed before the buildings were erected (i.e. provided there is no pre-let). Inmany cases this apparent loophole in the 1980 Act (meant to cover situations where local authorities lease existing buildings for their own occupation or for use as schools, etc.) would not be of much use as the developer needs the strength of the local authority covenant before construction starts. Rental guarantees on the other hand are not pre-lets and do not count as Prescribed Expenditure, but if the guarantee is called it would appear that notional capital expenditure would then be incurred, making this a potentially dangerous approach.

4. DEFERRED PURCHASE OR COVENANT SCHEMES

An alternative and possibly more flexible arrangement than the lease/leaseback is the deferred purchase scheme which again was mentioned in an earlier section. 'Covenant' or deferred purchase schemes involve a merchant bank developing property for a local authority, or appointing the authority as agent to carry out the scheme itself, but without any legal interest being granted by the authority, with the capital cost (including interest) usually being repaid over 5–10 years. This repayment period is often extended to up to thirty years to reduce annual payments, each of which (prior to July 1986) counted against that year's block allocation. The funder does not normally take an interest in the land (and the buildings) in these arrangements, instead relying entirely on the security and covenant of the local authority. As no lease was involved this method used to spread the capital cost of major schemes over the repayment period as only the annual payment counted against that year's block allocation. Since 22 July 1986 this is no longer the case and the total cost of the work now counts as prescribed expenditure in the year in which it occurs, regardless of when payments are made. Nevertheless, this arrangement is beneficial to local authorities who have a surplus of authority to spend but little cash (due to rate capping).

In March 1987 the DoE announced that occasional one-off projects up to three million pounds in cost, beginning in any one period of five consecutive years, was exempt from this new regulation and only the annual repayments would count as Prescribed Expenditure in the year incurred. Even where this exemption does not apply and the full cost of construction counts as Prescribed Expenditure, no borrowing is deemed to occur so deferred purchase is still a useful means to finance development providing there is some way of making up the shortfall between the 30% notional capital received when the development is let and the capital expenditure incurred. It would appear that the back-to-back provision contained in Circular 5/87 would not apply as the capital expenditure is not being used to acquire an interest in land. For deferred purchase schemes interest rates are normally slightly higher (about ½% above LIBOR) than for finance leases as the funder receives no capital allowances because no legal interest is transferred. But restrictions on choice of tenant, early redemption and options for tenants to purchase will not apply.

5. ADVANCE PAYMENT SCHEMES

Advance payment or prefunding schemes are the opposite of deferred purchase. Prior to July 1986 they allowed local authorities to spend against projects in advance of immediate requirement – for example, near the end of a financial year if there was likely to be an underspend – and so benefit in subsequent years from the generated surplus capital allocation capacity which could in turn be used to finance further projects. This used to be applicable where one year's allocation was likely to be underspent, where capital receipts would not all be spent or where a local authority envisaged that the regulations were about to change. Recently, for example, the retention limit on capital receipts was reduced from 50% to 30% (except housing receipts, which were reduced from 40% to 20%). Since July 1986 however the same regulations apply to advance purchase schemes as apply to deferred purchase schemes.

Non-Prescribed Expenditure without borrowing

Although expenditure on economic development is generally considered to be Prescribed Expenditure there are ways of achieving this spending without it counting against an authority's spending allocation.

Expenditure by a local authority's own pension fund may be one such way although if investment occurs inside the local authority area uncertainty exists as to whether this would count as Prescribed Expenditure. The 25% limit on property investment by local authority pension funds was abolished in 1983 but as the pension fund trustees have a duty to invest wisely and achieve the best possible return for the benefit of future pensioners, it is therefore arguable whether investment in marginally profitable or low return industrial investment in inner city areas would qualify.

Expenditure by economic development companies set up by local authorities, but otherwise supposedly independent from them, will not count as Prescribed Expenditure. By mid-1987 there were estimated to be more than 70 examples of development companies many of which have been formed to undertake specific projects, a few of which are enterprise boards with longer-term, more wide-ranging, interventionist objectives (The Planning Exchange, 1987). Advantages include a more streamlined and flexible decision-making structure free of local authority statutory controls, with less public accountability and democratic control. As they are independent companies they can be marketed as separated from the local authority which has advantages in terms of image with the private sector. They can be more single-minded in dealing with problems and staff can more easily work as a co-ordinated team without day-to-day political pressure. However, there are drawbacks as such companies are liable to corporation tax, VAT and the general requirements of the Companies' Acts. They must soon also be seen to be quite separate from the local authority as a June 1988 consultation paper from the DoE emphasizes, otherwise if they are controlled or dominantly influenced by local authorities, their capital spending will be brought within the system of capital control (so the consultation paper recommends). If these guidelines are strictly adhered to they may defeat their original objective (or lead to restructuring of the company involving 3rd party input to continue to achieve the objectives).

Barter arrangements were another possibility and have been used by some authorities to promote development, but the March 1988 clampdown has brought barter deals within the capital control system and they now count as Prescribed Expenditure. Typically the authority would exchange a freehold site in its ownership for a freehold building at an agreed value. Although these barter deals appear to be direct swaps and therefore do not involve additional expenditure, Nicholas Ridley (S.O.S. for Environment) considered they affected expenditure as there was the opportunity cost of foregoing a cash receipt which could be used to reduce debt. An alternative type of barter transaction – and no doubt the cause of the 1988 clampdown – was where one company took a lease of an existing asset for a premium subject to the local authority taking a leaseback at a rent. The premium was not actually received by the local authority but was used by the company to build the required development. Such arrangements were only possible, of course, where the authority had existing assets.

The receipt of planning gain can also be used to achieve economic development, whether or not the land is in local authority ownership. This, of course, depends on the local authority being in a situation where applications for significantly sized, profitable development occurs. In many of the less prosperous parts of the country this may be unlikely. Again the March 1988 regulations have thrown doubt as to whether planning gain will count as Prescribed Expenditure and so be brought within the capital control system.

12.5 Changes to the system of control of capital expenditure

The Green Paper, Paying for Local Government (January 1986) and subsequent consultation paper (February 1986) proposed a number of changes to the present system of controlling capital expenditure by local authorities and in particular to further restrict their freedom to use capital receipts because the volume of receipts accumulated from previous years (the cascade effect) was seen to enable some authorities significantly to increase their expenditure beyond the government's desired limits (even though this was really only re-cycling assets). A further problem was the difficulty for the government of predicting the level of receipts in any year. They were also worried that the 30% rule, introduced to limit the amount of extra spending, was at the same time a disincentive to sell public sector assets, another government objective.

Their main proposal, subsequently abandoned before the June 1987 election, (because of adverse reactions from local authorities of all political colour) was to remove the ability to spend receipts when they occurred and accumulated receipts (which mount up over time due to the cascade effect). The intention was to replace this system with one where local authorities had the ability to spend a percentage of receipts received over the previous three years only. Authorities would have been required to complete returns to government, giving information about capital receipts. The total spending limit (i.e. gross capital expenditure) for each local authority would therefore have *included* an assumed level of spending from capital receipts (i.e. a percentage of the previous three years' receipts) whereas, at present, spending limits (i.e. net capital expenditure) can be increased by capital receipts thereby giving some flexibility to authorities. In addition, the preferred system would have continued to permit revenue contributions to supplement capital expenditure but only to a limited, specified, extent and spending limits over a two to three year period would have been given, rather than one year at a time as at present. This was intended to respond to the criticism made by the Audit Commission in 1985 who argued for a three to four year planning horizon to enable forward planning and the spreading of the cost of capital projects over more than one year.

The system proposed in the 1986 paper but later abandoned was therefore one of controlling gross capital expenditure and the borrowing limit for each authority would have been less than, or equal to, their cash limit *less* the assumed level of capital receipts and revenue. The objective of this system was to exert greater control over capital expenditure and borrowing, enable responsible capital expenditure planning but at the same time encourage asset sales. To ensure more effective overall control on spending a number of detailed proposals were suggested to eliminate ways round the controls as well as the major change on capital receipts mentioned above.

These detailed proposals included bringing short leasehold acquisition (less than 20 years) into the definition of Prescribed Expenditure and ending advance purchase agreements and limiting the use of deferred purchase agreements. The latter proposals were in fact implemented later in July 1986 and the former proposal formed part of the March 1988 clampdown on leasebacks and barter deals (The Local Government (Prescribed Expenditure) (Amendment) Regulations 1988 and the local Government Finance Act 1988). The use of separate companies by local authorities was to be encouraged where it was appropriate for the scheme to be carried out, e.g. a partnership between the public and private sectors, but where such companies were set up simply as a means of circumventing public expenditure controls this was to be discouraged but no means of discouragement were mentioned. Once again, although the consultation paper was abandoned, new specific regulations bringing capital expenditure by companies controlled or dominantly influenced by local authorities within the system of controls applying to councils was proposed in a 1988 Consultation Paper devoted entirely to this subject (DoE, June 1988).

Perhaps the most interesting proposal in the 1986 Consultation Paper, and one more relevant to the theme of this book, concerned schemes where expenditure matched receipts – typically where a local authority exercises its compulsory purchase powers in order to sell land for immediate disposal to a developer or takes a head lease on an industrial estate and then sublets the units. The problem of Prescribed Expenditure not being matched by capital receipts in the year in which expenditure is incurred was discussed at length above, but this problem would have been exacerbated by the proposed new system where a local authority would normally have no spending power from a receipt in the year of disposal. The government proposal was therefore that where expenditure and disposal occurred in the same financial year local authorities should not be assumed to have incurred Prescribed Expenditure and where the proceeds simply offset expenditure they would not be treated as capital receipts for the purpose of generating additional spending power in future years. Fortunately, although this review of the local authority capital expenditure system was not implemented (or put to one side temporarily) this modification to the present system was implemented under Circular 5/87.

Gradually most of the methods discussed earlier of bypassing central government controls on capital expenditure have been brought within the control system. These controls are a major problem for inner city industrial development in particular, where schemes are marginal in profit terms and make direct development by local authorities a difficult option. A possible exception is where authorities have significant land and property holdings and therefore the potential for a succession of capital receipts to fund a programme of development over a number of years or in the fifty-seven areas that can put in IAPs under the revised urban programme. Yet even in these areas direct development is not common, mainly due to the limited amounts of capital available and the constraints imposed by the DoE. The most common ways of achieving development are by using finance leases, or lease/leaseback arrangements and deferred purchase schemes. Often private sector development partners are involved but where they are the local authority effectively almost guarantees the developer a return on capital by taking a head lease at a pre-arranged rent. All these methods now involve Prescribed Expenditure and even though the income received from occupational tenants counts as notional capital receipts, the 30% rule means that in certain situations they may only partly offset expenditure incurred. Circular 5/87 and back to back schemes have helped but to a limited extent as it is far too tightly constrained (Audit Commission, 1988). This resource mismatch was often a sufficient reason in itself to prevent an otherwise attractive development from proceeding unless a substantial volume of additional capital receipts could be achieved by sales of other assets owned by the authority. But it is this last point that worries the present government – the potential for local authorities to increase expenditure by selling assets and the amount of outstanding borrowings which are a significant proportion of national debt.

'over the years, councils have built up a very large stock of capital assets. Much of this was paid for out of borrowed money. Their borrowings are now about £45 billion, which is roughly 25% of the total national debt. Debt charges, which fall to ratepayers and tenants, amount to £6 billion a year.

Local authority capital expenditure and finance is a major component of the national economy. It has to be regulated as part of the Government's task of managing the economy as a whole.

During the 1980s, our control system has focused on the capital spending itself. But the system has been unsatisfactory for both central and local government. Everyone agrees that we need a change.' (DoE July 1988)

The change is proposed in a 1988 Consultation Paper aimed to take effect from 1990–91 to coincide with the introduction of the community charge or 'poll tax' and the new system of current finance.

Capital expenditure and finance after 1990

The July 1988 Consultation Paper considers that the 1980 system of controls over net capital expenditure had four deficiencies (although in 1980 Ministers stated that this system was more flexible and effective than the 1970 system).

1. It failed to ensure that capital expenditure was consistent with Government expenditure plans due to the amount of expenditure generated by capital receipts and the difficulty of forecasting their level (they have grown significantly, which ironically is a separate DoE objective). The overspend has been as high as 44% of planned gross expenditure;
2. The capital expenditure cash limit for each local authority does not take account of the capital receipts likely to be achieved by that authority, nor the proportion of past receipts available due to the cascade. It is probable therefore that some authorities have excessive spending power and others too little. Spending power from capital receipts has increased to £3.5 billion a year, which is greater than the rate at which new receipts are being realized. In 1987/88 capital receipts were forecast to be 53% of total local authority spending power (although some receipts will be used to repay debt but central government have no control over how much);
3. The 1980 system was circumvented by creative accountancy and loopholes in the legislation, but the most widely exploited loopholes have been closed by subsequent legislation (as described previously);
4. Owing to the frequent changes to the 1980 system, it has not provided a stable framework within which long-term capital programmes can be efficiently administered.

The proposed system (confirmed in the 1989 Local Government and Housing Bill) therefore has been devised to exert more effective control over expenditure and borrowing in aggregate and its distribution between areas and services. At the same time it aims to reduce the size of the public sector by asset sales and to provide the basis for longer-term planning. The new system will control credit to finance capital expenditure and ensure a reduction in local authority debt from asset sales. Capital expenditure will be financed in three ways:

(a) Borrowing or its equivalent. However money is obtained (except from

(b) and (c) below) it will be treated like borrowing as it has the same economic effect as borrowing. So, for example, lease and leaseback and deferred purchase schemes will count as being equivalent to borrowing;
(b) Government grants or contributions from third parties;
(c) Local authorities' own resources such as revenue contributions, profit from trading undertakings and capital receipts (the proportion not used to redeem debt or to be set aside to meet future commitments).

In essence the new system is in many ways similar to the 1981 system (as modified by recent legislation). Government will give each local authority a credit approval limit (covering (a) above) which is in effect similar to the 1980 spending limit. This limit will cover the next financial year and an indication of the minimum credit approval for the following two years. Credit approvals may be increased by supplementary approvals covering particular projects or programmes.

Capital expenditure above the limits implied by the credit or borrowing limit can occur through (b) and (c) above. Where capital receipts are used there is a significant change from the 1980 system. The proposal is for a 50% limit on the use of capital receipts for new capital investment, the remainder is to be used for debt redemption or to be set aside to meet future capital commitments (in which case future credit approval will be reduced) or as a substitute for future borrowing. This limit can be varied by the Secretary of State and the limit for housing receipts is intended to be only 25%. These restrictions on the use of capital receipts have been set taking account of accumulated capital receipts under the 1981 system (the cascade effect) but they do not apply where property, occupied by a local authority for a particular purpose, is to be sold and replaced by other property for the same purpose.

Under the new system the Government will be able to take account of individual local authorities' ability to raise money through capital receipts when setting the credit approval limit for each local authority (not possible under the 1980 system). This is a two-edged sword and whilst it may assist some authorities with few capital assets it may mean other authorities having low credit approval limits taking account of accumulated past capital receipts and potential future receipts. This could easily mean a significant cut in many local authorities capital programmes due to (i) the end of the cascade system meaning less total spending power from capital receipts, (ii) some authorities selling fewer assets than the DoE calculates in setting their credit approval limits and (iii) the absence of tolerance between years which will therefore encourage authorities to play safe and underspend to avoid being *ultra vires*.

Nevertheless the Consultation Paper is of the opinion that the new system is more flexible and more advantageous to local authorities than the system proposed in 1986, but with only 50% maximum of capital receipts being available to fund new capital spending it seems to fall some way short of meeting the criticism that 'Central Government . . . ought to find means of providing local authorities with greater incentive to realize capital receipts from surplus property holdings.' (Audit Commission, 1988.)

Conclusion

There is a long and logical history of direct local authority involvement in property development. It enables authorities to exert greater control over the development process and to benefit from the financial rewards of property development, depending on the degree of risk they share. But probably the most important reason, in the 1980s particularly, is that in many areas they have no alternative if new development is to be initiated. This may be because of their compulsory land acquisition powers or because certain areas are initially 'no go' areas for the private sector without local authority involvement. In some cases this has forced direct development by the public sector with no private sector development involvement.

It is ironic that at a time when local authorities' spending is severely restricted (between 1978/79 and 1986/87 capital spending has fallen by 40%) and their ability to become involved in property development is limited, the problems of the inner city have increased, requiring greater public sector involvement to underwrite private sector development risks. It seems likely that this dichotomy will continue and that increased capital spending will be required, either to improve the environment/infrastructure in order to encourage private sector involvement, or to facilitate more direct inter-vention through partnership schemes and direct development.

The capital expenditure control system needs to be changed to enable this to happen. As in the past no doubt much time and effort will be devoted to lessening or finding ways round the system but whether this is the most profitable use of time is another matter.

> 'So long as there are problems caused by restrictions and limitations which are regarded as being unreasonable then there will be those who explore ways and means of minimizing the impact of these restrictions. If that is creative accounting, then its days cannot be numbered and it will thrive in a variety of forms, geared to the needs of the moment, until greater freedom of action is achieved through the use of the kind of automatic checks and balances which maintained an acceptable equilibrium in the 'good old days'.'   (Blackburn, 1986.)

12.6 Legal powers to acquire, service, develop and dispose of land and to provide assistance to firms

Local authorities have a wide range of legislative powers which permit eco-nomic development enabling them to acquire, service and dispose of land, provide environmental improvements, develop, redevelop, repair or improve land or buildings, and support the private sector in the same activities through grants, loans and guarantees or by entering into partnership arrangements of one sort or another. Not all local authorities are fully aware of the scope of enabling legislation in part due to rather vague legislation not specific to economic development. The Local Government (Miscellaneous Provisions) Act (1982) helped to clarify this situation so that, whilst there may be financial reasons limiting local authority involvement, there should no longer be legal reasons to limit it. Nevertheless central government consider that:

> 'The present position is clearly unsatisfactory. An authority's ability to spend on economic development is governed not by properly focused powers or an objective assessment of what is needed and likely to be cost effective, but by the detailed interpretation of an assortment of powers that are, or are perceived to be, available. This is not conducive to sensible decision making on economic development issues.' (DoE, July 1988: Conduct of Local Authority Business, para. 7.12.)

Legislation is therefore planned to give all local authorities a new specific power to carry out economic development, although the extent of this power is not spelt out in the Government's white paper response to the Widdicombe Report other than saying that it will be 'circumscribed' that it will be financially constrained (although in a written parliamentary answer, at the end of 1988, the Local Government Minister stated that there would be no general financial limit on the use of this power as the general financial discipline to which local authorities are subject will be sufficient) and that local authorities will be discouraged from becoming involved in activities which the private sector or other public sector agencies can undertake 'more appropriately'. What it will also mean is that the ability to use other specific powers for economic development and Section 137 of the 1972 Local Government Act will be removed.

Section 137 powers will still remain for other activities of a non-economic development nature which benefit an area and these will be made clearer and a new financial limit introduced to replace the 2p rate due to the replacement of rates by the poll tax. This will be set at a lower level reflecting the new separate economic development powers and will be population based tying in with the new poll tax or community charge. A limit of £5 per adult in areas with a single tier of local authority is mentioned (and £2.50 per adult for each authority in two tier areas).

Until these new proposals become law it is worth examining the existing enabling and specific powers which local authorities have used to undertake economic development and to consider whether the Government's criticism of the present situation is justified.

Acquisition of land

The London Government Act 1963 gave London Boroughs powers to purchase land by agreement for the benefit, improvement or development of the borough (Sch. 4.s.20), but the important act for all authorities is the Local Government Act 1972. S.120 gives powers to local authorities to purchase land, by agreement, for any of their functions or for 'the benefit, improvement or development of their area', although this land need not be within their area. Furthermore S.111 gives a general and very wide-ranging discretionary power 'to do anything (whether or not involving the expenditure borrowing or lending of money or the acquisition or disposal of any property or rights) which is calculated to facilitate, or is conducive or incidental to, the discharge of any of their functions.' This gives all authorities the power to form development or joint development companies in conjunction with other specific powers (see below for discussion in more detail). S.137 gives power to incur expenditure up to the value of a 2p rate product 'in the interests of their area, or part of it, or all or some of its inhabitants, so long as they are not authorized or required to make any payment by or by virtue of any other enactment'. The Widdicombe Committee's research showed that in 1984/85 economic development activities accounted for more than two-thirds of all Section 137 spending.

The 1972 Act gives general powers over land acquisition by agreement and is therefore relevant to economic development. Where specific Acts, e.g. Education or Highway Acts confer specific functions on authorities they normally confer specific powers of land acquisition as well and it is usual for these specific powers to be used. Another Act conferring powers relevant to economic development is the Town and Country Planning Act 1971. Section 112 as amended by S.91 of the Local Government, Planning and Land Act 1980 gives power to acquire land for industrial and commercial development by agreement or compulsorily, provided that the consent of the Secretary of State is obtained and the land is required for the purpose of providing for the relocation of industry. S.112 (1) sets out the grounds where confirmation of a CPO to acquire land may be given by the Secretary of State.

'(a) That the land is required in order to secure the treatment as a whole, by development, redevelopment or improvement, or partly by one and partly by another method, of the land or of any area in which the land is situated; or

(b) That it is expedient in the public interest that the land should be held together with the land so required; or

(c) That the land is required for development or redevelopment, or both, as a whole for the purpose of providing for the relocation of populations or industry or the replacement of open space in the course of the redevelopment or improvement, or both, of another area as a whole; or

(d) That it is expedient to acquire the land immediately for a purpose which it is necessary to achieve in the interests of the proper planning of an area in which land is situated.'

Some of these terms are vague and certain circulars have been issued to clarify their meaning. Circular 26/77 for example stated that the Secretary of State must be satisfied that the land is required in order to secure the treatment of the *area as a whole* but that in respect of land to be acquired in the interest of the proper planning of an area, the land must be needed *immediately* but not necessarily as part of a comprehensive operation. Although it is the local authority who propose acquiring the land, it can achieve the purpose mentioned in any way that it wants, i.e. through development itself or in partnership with the private sector or by the private sector exclusively. In considering whether the land is suitable for development, redevelopment, or improvement, the authority must have regard to approved or adopted structure or local plans, any planning permission in force and any other material planning considerations.

Where land is acquired compulsorily the purchase price is governed by the rules contained in the Compulsory Purchase and Land Compensation Acts. Where land is acquired by agreement, even though the authority had the power to acquire compulsorily, compensation would be negotiated and settled on the same basis. Where no such powers exist and authorities can only acquire by agreement, then negotiations will not be regulated in any way.

Finally, under S.122 of the Town and Country Planning Act 1971, as amended by Sch.23 of the Local Government, Planning and Land Act 1980, land held for planning purposes can be appropriated to other uses. So, for example, an authority could transfer land from industrial development to leisure development.

Development and disposal

S.124 of the Town and Country Planning Act 1971 gives a general power, subject to the consent of the Secretary of State 'to erect, construct or carry out any building or work on any land . . . which has been acquired for planning purposes'. As the Sheaf Report (Report of Working Party on Local Authority/Private Enterprise Partnership Schemes, DoE 1972) states, this enables authorities to use this section 'for the purpose of laying out an industrial estate or redeveloping a shopping centre' (p. 45).

Section 2 of the Local Authorities (Land) Act 1963 gives similar powers to the above Act and enables authorities to acquire land, carry out works on land and undertake development for the benefit of improvement of their area.

Many local authorities had specific Local Acts to give them additional powers for economic development covering land acquisition and development. Between 1963 and 1976, for example, 31 authorities obtained powers to carry out work to prepare or improve a site of any industrial building or for improvement or provision of services and facilities (Minns and Thornley, 1978). An earlier example was the Oxford Corporation Act (1933) which stated in S.49 (1)(a) that the authority could 'with the consent of the Minister lay out and develop any such lands and on any such lands may erect and maintain houses, shops, offices, warehouses and other buildings'. However in order to rationalize the large amount of local legislation which developed in a haphazard fashion over many years, all pre-1974 local Acts ceased to have effect by the end of 1986 unless the Minister postponed or exempted a particular local provision from repeal. Amendments to the Local Government Act (1972) and Local Authority (Land) Act (1963) by the Local Government (Miscellaneous Provisions) Act 1982 following on from the Burns Committee of Enquiry and subsequent DoE Consultation Papers have widened their scope to give all authorities sufficient powers to acquire and develop land.

Local authorities have power to dispose of land conferred on them by Sections 123 of the Local Government Act 1972 and Section 123 of the Town and Country Planning Act 1971 both as amended by the Local Government, Planning and Land Act 1980, Schedule 23. Unless a short tenancy is being granted when rent-free or reduced-rent periods can be used, disposal (letting or sale) must be at the best consideration that can reasonably be obtained having regard to all the circumstances unless ministerial consent has been obtained. This gives authorities more scope perhaps than is immediately apparent to determine the level of rent where a lease is being granted as the lease can contain user clauses and restrictions if appropriate in the circumstances so that the market rent is restricted.

Under Section 95 of the Local Government, Planning and Land Act local authorities are required to maintain a public register of land owned by certain public bodies in their area, including the local authority itself, where the land in the Secretary of State's opinion is not being used, or sufficiently used for the functions of the body concerned. More importantly Section 98 of the Act gives the Secretary of State power to direct a local authority to dispose of land. Before issuing such a direction he must give notice of his intention to do so and give the owner 42 days to make representations as to why a direction should not be made.

Directions have been issued to sell by auction, without reserve and without previously determining the planning position (Searle, 1987) and the growing use of this power (19 sites covering 48 hectares in August 1987 alone) is of increasing concern to local authorities acting as a powerful incentive to think positively about land in their ownership and ensure either disposal (freehold or leasehold) on their own terms or their initiation and involvement in a development. A recent appeal case (7 April, 1987) between Manchester City Council and the Secretary of State for the Environment

(Muir Watt, 1987) highlights these issues and whether a S.98 notice of disposal can contravene S.123 of the Local Government Act. In this case, the first of its kind, the local authority were directed to sell various parcels of land within the City amounting to a total of 10 acres within four months by auction without a reserve and without any right of the authority to bid. The authority opposed this direction arguing that the land market was depressed and prices realized would be less than if the land was retained until market conditions improved. In effect, the authority argued that by selling within four months they would be disposing of land for less than the best that could reasonably have been obtained. It was held that a delay in selling would be contrary to the objective of the S.98 directive and so, in the circumstances, sale by auction would produce the best possible price at that time. The appeal by Manchester City Council was dismissed and an application for leave to appeal to the House of Lords was refused.

Powers governing assistance to the private sector

Sections 3 and 4 of the Local Authorities (Land) Act 1963 give all authorities the power to give loans for the erection of buildings or in pursuit of building agreements. The Local Government (Miscellaneous Provisions) Act 1982 enables such loans to be up to 90% of their valued security (previously 75%) whether this is the land itself or other assets owned by the applicant. S.137 of the Local Government Act 1972 as amended by S.44 of the 1982 Act also enables expenditure to be used to give financial assistance to the private sector carrying on commercial or industrial undertakings by way of loans, guarantees or grants, even though the 1982 Act did not increase the total amount that could be spent (product of a 2p rate per annum). For designated districts under the Inner Urban Areas Act 1978 powers to give assistance to the private sector are increased so that long-term loans for up to thirty years can be given for land acquisition or development for employment purposes which are in the interests of the area. The land need not be owned by the authority and there is no restriction on the uses to which the land or buildings can be put. As mentioned above these loans can be for up to 90% of the mortgaged security. Loans or grants can be made to help establish co-operative enterprises and also for environmental improvements in Industrial Improvement Areas set up under the Act. Within these areas grants can be given to firms for conversion, improvement or extension of buildings for, or to, industrial or commercial use up to 50% of the cost of works. Partnership Authorities have even greater powers including the waiving of interest for two years on loans for specified clearance and site servicing works, grants to reduce the rent on commercial or industrial premises and grants to lessen the interest payments on land and buildings to small firms (less than 50 employees).

Where land is derelict, substantial grants are available to the public and private sectors from central government (see Chapter 9, sections 9.3 and 9.4 for further details). Grants to the private sector used to be channelled through the Derelict Land Grant, but are now channelled through the City Grant which is funded totally by central government with no local authority contribution. Public Sector Derelict Land Grant is still available to local authorities.

Development companies

Section 111 of the Local Government Act, 1972, provides local authorities with a wide ranging power to intervene in the local economy. As mentioned earlier, one way permitted is by setting up companies which can be funded, at least initially, out of the product of a 2p rate per annum (S.137). Such companies have been set up by many local authorities, particularly the larger ones, where they are often called Enterprise Boards due to their wide-ranging interventionary activities. One purpose of some of these companies has been to bypass the expenditure controls referred to earlier in this chapter (but see below). Additionally such companies provide a way of utilizing money allocated but not spent by the local authority in any financial year and achieve a similar effect to schemes of advanced purchase referred to earlier, but without the adverse consequences, recently imposed by central government, that when this money is spent in subsequent years it will count against the capital spending allocation in the years in which it is actually spent. However, there is a limitation on the amount that can be paid into a company in this way.

These companies have had much greater flexibility of operation as they have not been so constrained administratively or as to the type of project that could be undertaken as their local authority parent. But they are liable to tax and where S.137 expenditure is used 'the use of such resources by the company or fund must be limited, by legal agreement with the authority, to projects which the local authority itself has the power to undertake. The fund is simply a device to carry out such projects in a different fashion' (Allen, 1984). The company or fund should be able to carry out such projects more quickly, more decisively, more efficiently and more effectively.

The 1988 DoE Consultation Paper on Local Authorities' Interests in companies contains the result of a survey which shows the extent of companies with a local authority involvement. 45% of all authorities had an interest in separate companies which in total amounted to 470 companies. In about 25% of all such companies more than one authority was involved and in over 60% of companies the interests held by local authorities were minority interests. By far the most common purpose of local authority companies is economic development (about a third of all companies or nearly a half if companies involved in training are also included). In well over half these companies local authorities only have minority interests. The majority of local authority companies either received no local authority support or less than 25% of turnover in the latest year for which figures were available. Of those companies that received over half of their turnover from local authorities, 50% were not local authority controlled.

The Government generally favours the use of companies by local authorities as they are consistent with the desire to introduce sound management practice into the public sector and they are a useful means to encourage joint ventures with the private sector. What the Government does not like is that some companies have been established to provide a means of avoiding controls on capital spending and borrowing. The Consultation Paper therefore distinguishes between local authority controlled companies, local authority dominantly influenced companies and companies where local authorities hold a minority interest.

The Paper proposes that controlled companies should not be able to engage in activities in which the authority has no powers to engage (subject to a few minor modifications) and they should be treated as being part of the public sector for the purpose of controlling expenditure. All the controls which apply to local authorities would therefore apply to controlled companies. Tax arrangements would not alter and companies would continue to be liable for corporation tax and VAT.

Local authority dominantly influenced companies (e.g. where a company is set up by serving officers of the local authority or where the majority of business is associated with a local authority or where the authority have a dominant, but minority, interest in a company) will also be governed by the *ultra vires* rule unless the company 'seeks to do things not permitted to the authority, with no involvement on the authority's part:' The Government, however, 'proposes to restrict the types of the companies in which local authorities may hold a minority interest to those where a specific interest is thus served.' Examples provided (relevant to the subject of this book) are enterprise agencies, the main purpose of which is to promote economic development; a company managing the common parts of land or buildings where the local authority is landlord of one or more of the buildings; a company to which the local authority has sold land with a requirement to develop the land or where the local authority has entered into a planning agreement with the company to develop land, but in either case providing the company retains an interest in the land.

Local authority dominantly influenced companies will be treated in a similar way to controlled companies in respect of financial controls. So capital transactions will count against the appropriate control total for the local authority itself. Where a local authority has only a minority and no dominant interest in a company it will not count as part of the public sector and will not come within the system of controls over capital and current expenditure.

References

Allen, M. (1984) *Local authority economic development initiatives. The fiscal and financial framework.* In *Planning and Employment in London. Current Issues for Practice*, R.T.P.I., London Branch.

Association of London Authorities (1987) *Restoring Local Government Finance*, Assn. of London Authorities, London.

Association of District Councils (1983) *Economic Development by District Councils. Paper One. Financing Economic Development and Aid to Industry*, revised, Ass. of District Council, May.

Audit Commission (1985) *Capital Expenditure Controls in Local Government in England*, HMSO, London.

Audit Commission (1988) *Local Authority Property. A Management Handbook*, HMSO, London.

Barrett, S. and Boddy, M. (1979) *Local Government and the Industrial Development Process*, School for Advanced Urban Studies, Bristol University.

Blackburn, J. (1986) Assessing the impact of successive rounds of restraints. *Municipal Journal*, 20 June.

Cross, C. and Bailey, S. (1986) *Government Law*, 7th edn, Sweet and Maxwell, London.

DoE (1972) *Report of Working Party on Local Authority Private Enterprise Partnership Schemes.*

DoE (1986a) *Review of the Local Authority Capital Expenditure Control System in England and Wales.* A Consultation Paper, February.

DoE (1986b) News release, No. 541, 15 October.

DoE (1987a) News release, No. 133, 17 March.

DoE (1987b) Circulars 5/87 and 22/87 – Capital programmes, 25 March and 18 September.

DoE (1988a) *Local Authorities' Interest in Companies.* A Consultation Paper. June.

DoE (1988b) *Local Government in England and Wales. Capital Expenditure and Finance.* A Consultation Paper. July.

DoE (1988c) *The Conduct of Local Authority Business. The Government Response to the Report of the Widdicombe Committee of Inquiry*, July HMSO.

Douglas, I. and Lord, S. (1986) *Local Government Finance. A Practical Guide*, Local Government Information Unit, London.

Local Government Information Unit (1987) *City Swindles. Urban Policy in the Eighties.* Special Briefing No. 21, Local Government Information Unit, London.

Minns, R. and Thornley, J. (1978) *State Shareholding*, MacMillan, London.

The Planning Exchange (1987) *Economic Development Companies*, Local Economic Development Information Service (LEDIS), July, Planning Exchange, Glasgow.

Searle, B. (1987) *Valuation & Estate Management – Valuing your Assets.* Conference paper on The Effective Management of Land and Property Resources, 26 June, Royal Institution of Chartered Surveyors, London.

Smith, P. (1987) Financing local authority development. *Estates Gazette,* 9 May.

Ward, I. (1987) Time to overhaul our capital spending control policy. *Municipal Journal,* 27 March.

Watt, J. Muir (ed.) (1987) Estates Gazette law reports. *Estates Gazette,* 7 November.

Widdicombe (1986) *The Conduct of Local Authority Business,* HMSO, London.

Part Four
DEVELOPMENT APPRAISAL

13
Private sector land ownership

13.1 Introduction

The complexity of the property development process, involving numerous different professions and the expenditure of large amounts of money often over many years, means that this potentially risk-taking activity requires sophisticated methods of researching market demand and development feasibility before a project proceeds. This has not always been so and some would argue that perhaps not enough developers even now are prepared, or even able, to undertake such work, still relying on hunch or 'back of the envelope' calculations.

The aim of this chapter is first to consider market research and the development industry, to examine the different types of developer active in the industrial and business space market; to describe the factors which will affect the likely demand for a completed development and hence its marketability; and to show the major factors involved in a specific site appraisal. Secondly, to illustrate how financial appraisals are undertaken when the site to be developed is freehold (see Chapter 14 for leasehold development and non-private sector development) and lastly, to show how such appraisals can be refined and extended to provide more information and confront the prospect of inevitable uncertainty.

13.2 Market research and the development industry

There are a variety of different types of developer active in the property market and each has different objectives towards development and therefore approaches development appraisal and the assessment of risk in different ways. The days when the typical developer was a small independent property company have long since gone although small companies are still very active in the development market and when boom conditions occur there is a growth in the formation of new companies. Property companies occupy an important part of the market, in some cases acting as providers of long-term investments for the institutions and some of these companies might still correspond to Oliver Marriott's description of the typical developer:

> 'The developer is a pure entrepreneur. The only equipment he needs is a telephone, and there were powerful developers who operated from their study at home or from a one-roomed office with a secretary, from there they wielded the talents of various professions: estate agents, solicitors, bankers, architects, quantity surveyors, consulting engineers, building contractors, accountants. The end result was a building.' (Marriott 1969, p. 21.)

Such companies are invariably traders and inevitably have a short-term view limited to the development process itself; but they may have motives beyond pure financial reward. Whilst a few view the promotion of good architecture as a main objective, more commonly it is the excitement and risks associated with the development process, and the satisfaction of seeing their own finished buildings, that is the motivating force behind the smaller developer. The larger, older established, property companies are less likely to undertake new development on such a significant scale relative to their size, being more content to exploit their existing portfolios which have been acquired or developed over many years.

As already mentioned in Chapter 4, the financial institutions are not only a source of finance for development by property companies, subsequently acquiring the completed projects as long-term investments, they increasingly develop directly themselves using in-house staff or surveying firms/ property companies as project management consultants. They have been willing to take all or some of the risks associated with development, in view of the greater overall return and the greater control over the development process likely to be achieved, securing ultimately investments matching their long-term requirements. Due to their long-term view, the risks associated with speculative development are reduced compared to the small property company whose ownership will often cease on completion of development and as a result the institutions are often prepared to accept lower profit margins.

Building contractors, particularly the larger companies, are increasingly involved in development risk-taking in their own right, or in partnership with property companies, partly as a means of maintaining a constant stream of work and partly because of the profits to be made in a closely related activity. They will often operate on slim development profit margins as continuity of development work bolsters the returns from their constructions side. Similarly, some large retailers now have their own development companies who work independently of their parent companies (e.g. Dixon Commercial Properties, Chartwell (Woolworth)).

The traditional property company also faces competition from owner occupiers developing for their own occupation, particularly the large companies in the industrial sector who can raise finance internally and develop a building tailor-made to their own requirements. Such developments might even appear financially unviable using traditional financial appraisal techniques but may provide other benefits to the occupier/developer such as a more efficient business, generating greater profits, or the prestige associated with an impressive (but expensive) headquarters building.

Finally, the public sector (central or local government) is sometimes involved in direct development on a speculative basis, particularly small unit industrial/workspace schemes in the less prosperous parts of the country (Chapter 11). The motive here is providing jobs rather than profit.

In other areas where the private sector is more willing to be involved (as risks are less) partnership arrangements between local authorities and private sector developers are more common with a variety of financial arrangements to apportion risk and management control.

13.3 Demand assessment

Market research into the potential demand for a proposed development is now very important due to volatile market conditions, a reduction on supply restrictions as a result of government policies (Chapter 9), and the need to provide convincing proof to financiers (banks, insurance companies and pension funds), that the proposed building will not only let well initially but is in an area with good long-term growth prospects and is designed with a view to present and possible future operational needs (new technology, etc.). The larger the scheme, and the more specialized the potential market, the more important such research will be as development risks will generally be greater. Assessment of both occupational demand and investment demand (yields, value, etc.) will be needed to justify development.

Due to the lack of accurate, detailed and reliable data and the disaggregated nature of the industrial and business space property markets, with separate submarkets according to specific location, size and age of building, demand assessment in particular is not an easy task and like all forecasting in the world of property, requires the greatest of care in its interpretation.

The demand for industrial and business space will primarily be affected by the state of the national economy and how that is likely to change over the period of time under consideration. Demand is also influenced by local and regional factors such as the strength of the local economy; the soundness and diversity of the local economic base; the probable growth (or decline) in population; the growth in the service sector relative to the local economy generally; central/local government attitudes, restrictions or incentives; the transport infrastructure; the general environment (housing, schools, shops, recreational facilities, etc.); the level of rents/rates and the potential as a location for decentralization and future expansion. Some of these factors are relatively easy to ascertain, others are not, particularly at the local level.

Where a proposed development is large and the timescale of development considerable, more specific analysis and prediction is sometimes pursued. For example, a relatively unsophisticated approach would simply be to analyse for various dates in the past (probably census years) the ratios of industrial/office employment to total employment, and total employment to total population. The rate of change in these ratios could then be assessed, projected and applied to future population forecasts with modifications made to take account of changes in population age structure, socio/ economic composition of the population, expectations for the strength of the economy, and increases or decreases in unemployment in different sectors of the economy, etc. Forecast employment would then be converted into floor space, taking account of past and projected future changes in employment floor space ratios. The difference between projected floor space and existing stock will give a measure of probable demand for additional floor space over the time period being considered and can be expressed in annual terms as a take up or absorption rate. Such relatively crude forecasts would normally only be used as a guide to future demand for say a whole town, and would have to be supplemented by other research and specific site studies.

For example, particular attention would be paid to a town's communication with other centres of economic activity (e.g. the Thames Valley towns and their excellent access to London and Heathrow Airport by road and rail), the working and residential environment, and the type and size of companies already established and their potential for growth. Will increased demand mainly be in small, medium or large buildings? Will tenants want new buildings or will refurbished buildings be acceptable? Will demand be for new self-contained units on FRI (Full Repairing and Insuring) leases or for small units in managed workspace schemes housed in refurbished buildings?

13.4 Assessment of supply

Of equal importance to an assessment of likely future demand is a study of existing and probable future supply; the developer needs an appreciation of the overall picture to gauge whether there is a likely imbalance between supply and demand. A developer will want to know the amount of space currently available to let, its percentage of the total stock (which can be compared with minimum necessary vacancy rates for normal market fluidity), the size of units available, their age and specification, their quality, their rent, the services available and, very importantly, their specific location. Similarly, the amount of space under construction (taking account of space demolished), its location, specifiation, etc. and likely completion dates will need to be studied as well as space in the pipeline (with or without planning permission). This latter figure can inevitably only be a crude estimate.

Supply and demand can then be compared to assess the degree of under/over supply at a future date coinciding with the likely completion date of the proposal being considered. Additional checks can be made on:

1. The stock/flow relationship to see whether new developments over the next few months/years represent an abnormally high percentage of total stock;
2. The vacancy rate and how it has changed over the last few months/years;
3. Past rental growth which will be a measure of the underlying strength of demand relative to the supply of new developments.

13.5 Specific location and site appraisal

From the foregoing it is clear that a detailed site analysis is as equally important as an overall demand/supply analysis. Having justified a proposal in general terms it must be justified in specific terms.

What are the characteristics of the site itself and its immediate surroundings? Does the site enjoy a prominent position? Are the neighbouring buildings new or old, in good or bad condition, in mixed or industrial use? Are they likely to be redeveloped in the near future thus changing the appearance of the area or is the proposal being considered on a greenfield site and sufficiently large to create its own local environment (e.g. Stockley Park and the larger business parks).

In considering the site itself, the developer will pay particular attention to factors which will affect design and costs. How level is the site, what is the soil structure, how effective is drainage and are all services available? What shape is the site and will it enable maximum floor space to be provided in simple cost efficient rectangular buildings or will more elaborate, complex and more costly designs be necessary and if so how will this affect rental levels (if at all)? If existing buildings are to be converted what are the floor loadings, what state are the stuctures and the services in? What shape and size is the building and is there sufficient natural light? How much car parking will be permissible by the local planning authority and how much can be accommodated on the site and at what cost? Finally, are there any legal restrictions such as rights of way or light which would affect the size, shape, design or specific position of the proposed building?

In today's market detailed research is essential to justify proposals and to lessen development risks. The cavalier attitude to developmental appraisal prevalent in the past is no longer appropriate although still by no means unknown. Just as market research has become more sophisticated, so has financial evaluation as the following sections illustrate.

13.6 Financial appraisal – the basic approach

The method most frequently used for appraising the financial viability of development schemes, whether new developments or refurbishment of existing buildings, is generally known as the residual valuation. The basic concept is straightforward, difficulties arise not in the method itself but in estimating the amounts of the many variables that go into the valuation, and this is where the expertise of the valuer is so important.

Essentially a developer will use a residual valuation for three main purposes corresponding to some extent with the chronological sequence of the development process:

1. To calculate the maximum value of a development site which a developer is considering trying to acquire. This ceiling figure (allowing the developer a reasonable profit) can then be used where a site is to be sold by tender or auction. Alternatively, where a sale is by private treaty the residual valuation can be compared with the asking price and a decision made as to whether it is worthwhile for the site to be purchased and development undertaken. Of course the site will not necessarily be a cleared site in which case the appraisal will be undertaken to test the feasibility of redevelopment or conversion/refurbishment.
2. To calculate the expected profit from undertaking development where a site is already owned by the developer. The site value is treated as a known cost of development and the residual item of the calculation then becomes the amount of profit the development should yield. If reasonable in all the circumstances, this will encourage the developer to proceed with the proposed scheme, and if unreasonable it will lead to delay, redesign, abandonment or possibly the application for a grant (e.g. City Grant).
3. To calculate a cost ceiling for construction where land has been acquired (and is therefore a known cost) to ensure where possible that a reasonable profit margin is maintained. Whilst construction is underway, progress will be monitored so that if viability is threatened, alterations to the scheme can be considered before it is too late.

At later stages in the development process, where more specific and accurate information about the project is available, more detailed calculations can be undertaken. The length of the building period and the cost of the building can be estimated more accurately enabling more detailed cashflows to be calculated and more refined appraisals undertaken, as described in section 13.9 *et seq.*

It is important to remember that the appraisals which private sector developers use to assess the viability of development schemes, are concerned solely with financial factors taking account of values and costs to the developer. They are not feasibility studies, or cost benefit studies, which would consider a much wider range of costs incurred, and benefits received, by people other than the developer himself.

In its simplest form, when used to assess the development value of land, the residual valuation will estimate the maximum purchase price of a site by deducting the expected total costs of development, including an allowance to cover risk and profit, from the expected price that the completed development could be sold for in the market. The residual valuation could therefore be expressed in the form of a simple equation where the answer is the residue, or sum left over, after deducting costs of development from the value of development:

	Sale value of completed development		A
less	cost of development	B	
	profit allowance	C	B+C
equals	residue for purchasing land		D
i.e.	$D = A - (B+C)$		

Alternatively, if viability is to be calculated when land cost is a known item of expenditure, the residual profit can simply be shown as:

$$C = A - (B+D)$$

In certain situations viability may be marginal and grants may be necessary to encourage the private sector to proceed. In such cases the above expressions could be modified as follows:

$$B + C + D \leqslant A + Grant$$

It is important to remember that in a market economy where the supply of new buildings is a small percentage of the stock of existing buildings, the value of land is determined by what can be developed on that land and the value and cost of that development. Furthermore, the value of that development is not directly related to its cost of production, but is created by the interplay of market forces, namely the supply and demand for similar properties which determine market price. For industrial and business space property there will be a user market which will determine rental levels and an owner/investment market which will determine capital values. Even where the occupier and owner are the same, the capital value will still relate to potential rental value which the owner occupier would otherwise pay if an alternative property was to be occupied.

It is also important to remember that, although the residual valuation can be expressed as a simple equation, there may be considerable difficulty in accurately estimating the component parts of that equation. For example, what was simply referred to as the costs of development may encompass a variety of different elements such as demolition costs, building costs, costs of drainage and external services, landscaping and car parking costs, professional consultants' fees, finance costs (which will vary as interest rates alter and as construction and/or letting delays occur), letting and sale fees, etc. These items are all variables and accurately estimating their values over an undetermined and possibly lengthy development period is not an easy task. Furthermore, with so many variables in the equation, slight changes in a few of them will almost certainly result in a wide range of answers (and could easily eliminate the profit allowance), owing to the sensitivity of the residual valuation. This is discussed in more detail later in this chapter. Of equal importance is the size and content of the scheme to be valued. Until detailed planning permission has been obtained, this will also be a variable and much time and effort will be involved in formulating the optimum scheme.

For these reasons, the Lands Tribunal are very wary of this valuation approach unless there is no simpler method of valuation available. Great care must therefore always be taken in using residual valuations and, wherever possible, comparable market land price evidence should always be used, at least as a check. For urban schemes an accurate comparison is often not possible as projects vary in content, cost and value whereas in suburban and greenfield development an accurate comparison method of valuation should be possible.

13.7 Residual valuation

A developer wishes to estimate the value of a cleared site of 0.8 hectare located close to London which is for sale, with outline planning permission for 4000 m² gross of industrial space. After consideration of various layouts and designs and the type of space suitable an optimum scheme is finalized which the developer feels confident would receive detailed planning permission. In this instance it is felt that the total floor space contained in the outline planning permission could not be improved upon. From knowledge of the area and the scheme proposed, the developer, with consultancy advice, considers that all building work could commence in six months' time and should take nine months to complete.

Current building costs including services are
estimated by the quantity surveyor to be: £325 per m² gross

Current rental values are estimated by the letting
agent to be: £75 per m² gross

The developer considers it prudent, after consultation with his agent, to allow a six-month period to achieve a letting. A simplified time chart may therefore be drawn up to illustrate the development period.

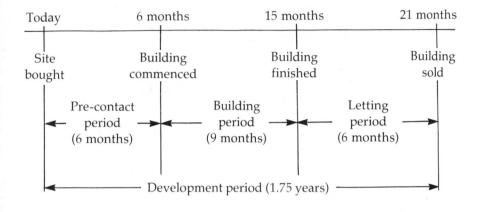

If the scheme goes ahead it could be funded from a mixture of internal finance (retained profits from previous developments) and short-term bank finance, until a decision is made about whether to keep the completed building as a long-term investment or to sell it and hopefully recoup a capital profit. This is discussed in more detail below (see also Chapter 4).

	£	£
Expected value of scheme		
Income: 4000 m² @ £75 per m²		300 000 p.a.
Yield @ say 8.25%		
(allowing for purchaser's costs)		12.12
Capital value		3 636 000
Costs of Scheme		
Building costs: 4000 m² gross		
@ £325 per m², including services	1 300 000	
Access roads, car parking, landscaping,		
etc. 4000 m² site area @ £35 per m²	140 000	
Demolition (assume cleared site)	—	
	1 440 000	
Professional fees and planning application costs		
Architect, Q.S. and other consultants		
@ say 10%	144 000	
	1 584 000	

		£	£
Contingencies @ say 5% of total costs incurred		79 000	
		1 663 000	
Short-term finance @ say 10% p.a. (compounded quarterly): On building costs, ancillary costs, professional fees and contingencies over half the building period of 9 months		63 000	
		1 726 000	
On total costs incurred on completion of building work for full length of letting delay (i.e. say 6 months)		87 000	
		1 813 000	

	£		
Letting and sale fees:			
Letting fee @ say 15% of income	45 000		
Advertising and marketing say	15 000		
Sale fee @ 2% of sale price	73 000	133 000	

Total development cost	1 946 000	
Return for risk and profit @ say 15% of capital value (or 17.6% of all costs including land)	545 000	
Total expected costs (incurred on completion of scheme)	2 491 000	2 491 000
Site value (in 1.75 years' time)		1 145 000
P.V. £1 1.75 years @ 10% (compounded quarterly)		0.8413
Less acquisition		963 000
costs @ say 2.5%		24 000
		939 000
Site value today	say	940 000

Alternative method of calculating site value today:

Site value (in 1.75 years' time)		£1 145 000
Let site value today be	£x	
plus acquisition costs @ say 2.5%	£0.025x	
Gross site cost	£1.025x	
Finance @ say 10% p.a. (compounded quarterly) over development period of 1.75 years	£0.1934x	
Total	£1.2154x which equals £1 145 000	

Therefore x = £942 000

Therefore *Site Value today* = £942 000

The slight difference between £942 000 and £939 000 is accounted for by the difference in the way the acquisition costs have been calculated. These costs should be calculated as a percentage of the actual site cost and not as a percentage of the gross site cost. £942 000 is therefore a more accurate estimation of site value, but as the difference between the two approaches is so small, either approach is acceptable.

Income

Although the buildings will not be available until completed in one and three quarters years' time, today's rental value is usually assumed to avoid the problems of predicting the future. However, in areas of strong demand, and hence competition for sites, where there is strong optimism for the future (perhaps based on good research), developers may project rents explicitly or take a view of likely future growth. This point is discussed in more detail later in this chapter.

Net rental value is used, which assumes that all outgoings such as internal and external repairs, insurance and rates have been allowed for so that the net income (before tax) is derived comparable to other non-property investments. It is normal practice today to ensure that the tenant is responsible for bearing these outgoings so that rental values are normally quoted on a net basis. If there are many tenants (as in workshop schemes, for example) then management costs to cover rent collection, and to ensure that repairs are undertaken, etc., should be deducted separately. Where a workshop scheme is fully serviced (Chapter 7) then the cost of these services must be deducted to produce a net income, calculated on the net usable floor space a tenant can occupy. For traditional factory/warehouse buildings usable floor space is taken as being equivalent to gross internal space. For high-tech/business space buildings which usually have a high office content there is an argument for deducting space which is non-usable in rental terms; such space would include stairs, lifts, landings, WCs, plantroom and normally entrance halls, but current practice is to treat high-tech/business space buildings as a type of industrial building and to quote rents on a gross internal basis. For small unit workshop schemes the net usable area which individual tenants occupy would be used to calculate rent.

Investment yield

It is common (but not universal) practice in the industrial/business space property market for there to be a separation between occupation and ownership of buildings. Unlike the residential market where owner occupation is common, many industrial/business space occupants prefer to pay rent rather than acquire property outright for a capital sum. In theory their capital can be more profitably employed in their business in which they are experts, rather than being tied up in owning property, at which they are unlikely to be experts. Consequently there is a separate investment market for property which is a (small) part of the overall general investment market. Investment yields in property must therefore relate to other investment yields. Even where property is owner occupied, capital value will often be calculated by applying an investment yield to the potential rental value, although sometimes comparable figures of capital value per square foot are used.

The yield has to reflect a number of factors, the most important being:

1. The state of the economy and the general level of interest rates, yields;
2. The location of the property and hence its future rental growth expectations;
3. The security of the income in terms of the tenants' continued ability to pay the rent;
4. The life of the property and the cost of obsolesence;
5. The number of tenants and the management bother attached to rent collection, supervising repairs, providing services, etc.;
6. The size of the investment (very large investments will have higher yields as fewer investors can afford the risk of having so many 'eggs in one basket' or the inflexibility if a smaller sum of money had to be realized by a quick sale).

The greater the rental growth expected the lower the initial return an investor would be prepared to expect. Conversely, an investment where income growth is expected to be small or even zero (such as government stocks or Building Society accounts, or secondary industrial estates) will have a high initial yield or return to compensate. Thus, undated government stocks (currently yielding about 9%) will give a fixed annual income of £9 for every £100 invested whilst £100 invested in prime high-tech business space property will (at the time of writing) give an initial annual income of about £7 (i.e. a yield of 7%). It is anticipated that the income to be derived from the property will increase significantly at every rent review so that in, say, five or ten years' time the annual income might exceed the income from government securities. It is part of the skill of the valuer to determine the appropriate yield, which can vary from below 4% for prime shops with good security of income, high rental growth expectations and little potential obsolescence, to over 15% for old industrial estate investments.

Purchaser's acquisition costs of about 2.5% (agents' fees at 1%, stamp duty at 1%, legal fees at 0.5% and VAT at 15% on agents and legal fees, possibly recoverable) would normally be deducted from the capital value to determine the net realizable value or alternatively (as in this example) this could be reflected in the investment yield used. If the development was being forward funded by an institution, stamp duty would only be payable on the cost of the site as the building is developed on behalf of the institution who therefore do not acquire it on completion.

Building costs

As stated earlier, various proposals will be costed before the most appropriate scheme is finalized. Expert quantity surveying advice is essential so that building costs are derived as accurately as possible, particularly as the major component of development expenditure is the payment to a builder for undertaking construction. Building costs are normally calculated on the gross internal area of the building(s), i.e. the area of the building obtained by measuring to the inside of the external walls. (This total area will be slightly less than the gross external area used for plot ratio purposes.)

It is common practice for this figure of building costs to be the tender price as at the date of undertaking the appraisal, i.e. the traditional developer's assumption is to use current rental values and current costs thereby avoiding the difficulty of predicting the future. A further justification for this approach is the assumption that increases in building costs will be matched by increases in rental values. Nevertheless, some developers prefer to use predicted building costs (which are considered to be easier to forecast than rental values) and provide a smaller allowance for risk and profit if present levels of rents are used. Forecasting and its effects on residual valuations is discussed later in this chapter.

VAT at 15% is chargeable on refurbishment costs (unless the building is listed or the front wall only is being retained) but currently (1988) new construction is zero rated, although due to EEC pressure a change will occur in April 1989 when all building work will be subject to VAT. Whether this will be recoverable is a complex problem beyond the scope of this book and will depend amongst other factors on what the developer elects and to whom the building is let.

A summary of typical building costs for different types of industrial and high-tech/business space developments are shown in Table 13.1 together with regional variation factors in Table 13.2.

Professional fees

These fees are usually based on building costs but are sometimes agreed as a negotiated sum. Except for small jobs, architects' scale fees for designing new buildings are 5–6%, although lower fees can sometimes be negotiated, quantity surveyors' fees are about 3% and, where necessary, structural engineers' about 2.5%. In addition, there may be other professionals involved in the design and construction, such as heating, lighting and par-

Table 13.1 Indicative costs/m² outer London, May 1988

[Competitive tenders in outer London for contracts let on a fluctuating basis, excluding external works, drainage and external services.]

Type of use	Cost/m² of gross internal area £
Factories	
for letting (incl. lighting, power and heating)	245–335
nursery units (incl. lighting, power and heating)	305–445
workshops	360–445
for owner occupation – light industrial use	315–445
Warehouses	
for letting, low bay (no heating)	185–225
for owner occupation (incl. heating)	270–325
for owner occupation, high bay (incl. heating)	365–415
High-tech office/industrial buildings	
for letting (shell, without raised floors, air conditioning or lift but with suspended ceilings)	335–485
for letting (raised floors, air conditioning and suspended ceilings)	545–715
for owner occupation (controlled environment, raised floors, suspended ceilings, lift, finishings and fittings)	705–910
Car parking	
surface level (tarmacadam, incl. lighting and drainage)	35
surface level (concrete interlocking blocks, incl. lighting, drainage and landscaping)	45

Source: Davis Langdon and Everest.

ticularly air conditioning experts, which will add further to the total fee. The size and complexity of the scheme and whether it is new-build or a conversion will determine the total amount to be added on for fees. VAT at 15% will be charged but normally can be reclaimed on new construction; however, the cost of submitting a planning application will add slightly to the cost of all development. For new industrial and business space schemes 10–13% would normally be allowed to cover these items, but for conversion/refurbishment the figures could be as high as 15%.

Contingencies

Some developers make allowance for contingencies as a separate item to cover the difficulties of precisely estimating building costs and the length of the building period; others do not, but provide a larger allowance for risk

Table 13.2 Regional variation factors

Region	Variation Factor
Inner London	1.10
Outer London	1.00
East Anglia	0.88
East Midlands	0.86
Northern	0.89
North-west	0.91
Scotland	0.87
South-east	0.93
South-west	0.86
Wales	0.90
West Midlands	0.85
Yorkshire and Humberside	0.86

Source: Davis Langdon and Everest.

and profit. Particularly where conversion or refurbishment of an existing building is being undertaken, it is advisable to provide a contingency sum as the cost of this work is very difficult to predict with unforeseen problems almost always occurring when building work is in progress. Similar unforeseen problems often occur when foundations are excavated even if trial boreholes are made before building work starts.

A contingency allowance between 3% and 5% of all buildings and related costs incurred is usual, the actual amount depending on what stage in the development process that the appraisal is being undertaken, the degree of uncertainty about the subsoil and particularly the structure of any existing buildings that are being retained and refurbished. Where the development scheme consists entirely of conversion or refurbishment then an allowance as high as 10% would normally be adopted.

Short-term finance

Development schemes involve considerable capital expenditure. It is normal practice for capital to be borrowed to finance this expenditure. The major sources for borrowing are the clearing and merchant banks, or the insurance companies and pension funds, who provide both short- and long-term finance and in some cases undertake development themselves (Chapter 4).

A developer works on the assumption that short-term money is borrowed to finance the cost of construction, which is then repaid at the end of the development period out of the proceeds from selling the completed scheme or from raising a long-term loan. As the money borrowed is a short-term loan the cost of borrowing is related to the cost of short-term loans generally, i.e. from say 1–6% above LIBOR (London Inter Bank Offered Rate), dependent on the status of the borrower, the size of the loan and the length of time that the loan will be outstanding, etc. At the time of writing (early 1988) LIBOR is 8.5–8.75% and a major development company might expect to pay slightly above this figure. Some developers might have secured preferential terms on longer-term loans in the past or might use internal funds generated from the profits of past developments. Nevertheless, if open market site value is being determined, the prevailing borrowing rate should be adopted as this reflects the opportunity cost of using this money. One exception to this is where a pension fund or insurance company might provide short-term money at a preferential rate of interest in return for obtaining ownership of the completed development at a preferential price (forward sale agreement).

It is normal practice for builders to be paid at monthly intervals for work completed – i.e. a proportion of the total building costs. Such payments are normally triggered by the issuing of a series of architects' certificates. Finance is raised accordingly and interest will therefore accrue in stages, although normally it will be rolled up and repaid as a lump sum on completion. So payments to the builder early on will involve interest accruing at a compound rate over virtually the full building period whereas payments made near the end of the building period will incur hardly any interest. It is therefore normal practice (certainly for an initial appraisal) to assume that the total amount is borrowed for approximately half the building period and the interest compounded quarterly. This means that a normal rate of interest, related to LIBOR, of say, 10% per annum, will be an effective cost to the developer of 10.38% per annum [i.e. $(1 + 0.10/4)^4 - 1$]. Allowing interest over half the building period also assumes that the individual monthly payments are symmetrical about the midpoint of the building period, i.e. that the total costs incurred during the first half of the building period approximately equate with those incurred during the last half. (This assumption is discussed later in section 13.9 *et seq.*)

Alternative methods of calculating short-term finance used by some developers are to allow interest over the full period but on half the building costs, or over the full period on total building costs but at half the interest rate. Each method will give a slightly different answer.

Professional fees are paid at intervals during the development period. The earliest (and largest) payments (to the architect particularly) are usually made before building work commences to cover the work required to obtain detailed planning permission, etc. Subsequent payments are then made at regular intervals during the building period based on a percentage of costs incurred. To be accurate therefore finance on these fees should be allowed for longer than half the building period due to the initial payments being proportionately larger than the remainder. In practice many developers lump fees together with building costs and allow for finance over half the building period on the combined sum as it makes the calculation easier and the margin of error is small as the fees are at most usually only 15% of building costs.

On completion of all building work there will normally be a delay whilst the building is marketed. Once tenants have been found, legal documentation and possibly rent-free, or fitting-out periods, will further delay the receipt of income. The length of this delay will obviously be dependent on the strength of demand at the relevant time and the rent that is quoted. In a poor location during a recession the delay may be considerable. Conversely, in a prime location when demand is strong, letting may be achieved while building work is proceeding and therefore no delay will be encountered, even though unfinished buildings are not so easy to let. To reduce risk, a pre-let may be arranged which will also eliminate delay at this stage of the development when maximum costs have been incurred, although the developer would expect to confer some rental concessions to the tenants. In some cases development will only commence *after* a pre-letting has been secured.

It is normal practice to make allowance of a few months letting delay, dependent on the location, and therefore finance will be required for the full period of the delay, as the total costs (building costs, fees and finance incurred during the building period) will be outstanding until the building can be sold, or refinanced, which will not normally be until the building has been let. This letting delay could be added to half the building period, and all the short term finance calculated in one stage, i.e. in this example interest at 10% per annum (compounded quarterly) for 10.5 months on total building costs and fees. Whilst the answer is obviously the same it is, however, often useful to separate the actual cost of delay to emphasize its magnitude and significance.

If the building is still unlet after say, six months, and the scheme has been foward sold, then in many cases the funding agreement will stipulate that the developer will be responsible for paying the rent for a further specified period or until an occupying tenant is found. In this situation the prospective profit margin will be eroded by rental payments rather than accumulating interest charges.

Letting and sale fees

Letting and sale fees are usually paid at the end of the development period and so finance will not normally have to be arranged to cover payment of these items. Letting fees are normally based on a scale of 10% of the first year's rent where one estate agent is involved or 15% where two agents are used (common practice for larger developments). Sometimes, on very large or easy to let buildings, a negotiated sum may be arranged below the level of the scale fee. The cost of advertizing and the preparation of brochures, etc. would normally be charged as a separate item.

Allowance for sale fees will depend on whether the development will be sold or retained as an investment. If the developer intends retaining the development as a long-term investment then no sale fees as such will be paid as no sale occurs, but, if long-term funding is necessary, then funding fees will be incurred and it is normal practice for the developer to pay both sides legal and agents' fees. A normal allowance for sale fees would be 2–2.5% of the sale price (i.e. agents' fees 1.5%, legal fees 0.5% and 15% VAT, which might be recoverable) but on a very large scheme the percentage fee may be smaller as the capital value would be much larger, but the extra work involved (still one sale) would not be correspondingly greater. Similarly, if one firm was involved throughout the development in finding the site, letting and subsequently selling the completed development, reduced fees might be negotiated.

Return for risk and profit

Property development involves the taking of risks which can be substantial where that development is of a speculative nature. Not only will costs almost certainly alter during the development period but rental levels, investment yields and the time that it will take to let (and maybe sell) are difficult to anticipate several years into the future. This is dramatically illustrated by the boom and crash of the early 1970s. A developer will therefore incorporate into an appraisal an allowance for those risks and also to provide a profit, or return, for the time and effort involved in creating an asset. The amount of this allowance will obviously depend on many factors such as the type of developer, the size of the scheme, the length of the development period, the degree of competition (and hence optimism of the future) whether costs (and rents) have been projected and whether the scheme is pre-let or forward sold. The longer the development period the more uncertain the future and therefore (other things being equal) the greater the allowance usually made.

This allowance might be expressed as a percentage or mark up on total development costs or as a percentage of the capital value created. As one item of development cost is the cost of the land and the object of the calculation is to find this figure, it is easier, and therefore common practice, to express the profit allowance as a percentage of capital value. A typical allowance for a speculative scheme without pre-lets, etc. would currently be 10–16% of capital value (12.5–20% of total costs including land), although, depending on the size and duration of the development, lower or higher figures could be appropriate.

It can easily be shown that there is a close relationship between the two forms of percentage allowance, whether expressed of capital value or of total costs, due to the fact that development costs, plus land cost, plus profit must equal capital value. So for example if the profit allowance is 20% of capital value, then it must also be equal to 25% of total costs. In like manner a 15% mark up on total costs is equal to approximately 13% of capital value, etc.

Residual site value and finance on site cost

The maximum value for the site that the developer could afford to pay is calculated by deducting costs and a profit allowance from capital value. This indicates the site value at the date the development is completed (i.e. the date when the development is let and capable of being sold at the capital value calculated). As the site has to be purchased and paid for before development commences it is necessary to calculate the site value at today's date.

Finance will be required to cover the cost of acquisition until the capital value of the completed development can be realized and all borrowing repaid. It is normal practice to allow for interest at the same short-term rate as adopted previously and to allow for it over the full development period since the full amount of the loan will be outstanding for the entire duration of the scheme. One and three quarter years has been taken in this example to include six months from the time the site is acquired until building work could commence. During this lead in period the design and layout of the building will have to be settled, detailed drawings and other building contract material prepared and a contractor selected. In most cases, detailed planning permission, or approval under reserved matters must be secured and Building Regulations approval obtained.

Although the calculation of the site value of £940 000 in the residual valuation shown above looks detailed and scientific great care must be taken in

Table 13.3 Land values over a range of ERVs and yields (assuming profit on cost of 17.65% and building cost per sq metre of £325)

Yield	ERV £/m²				
	70	72.5	75	77.5	80
8.0	850 000	933 000	1 017 000	1 101 000	1 184 000
8.125	813 000	895 000	978 000	1 060 000	1 142 000
8.25	778 000	859 000	940 000	1 020 000	1 101 000
8.375	743 000	823 000	903 000	982 000	1 062 000
8.5	710 000	788 000	867 000	945 000	1 024 000

using this method of valuation as stated earlier. Small errors or inaccurate estimates of any of the key variables can have a disproportionate effect on the residual answer. For example, if the rent actually achieved was £72.5/m^2 rather than £75/m^2 the site value would reduce by nearly £100000. Similar affects will occur for small changes in the other key variables of investment yield and building cost (see Table 13.3 above). Some form of sensitivity analysis should therefore always be undertaken as discussed in Section 13.11. (See also Morley (1988) for a more detailed examination of this point.)

However, before undertaking a full sensitivity analysis as described in section 13.11, a developer would probably undertake a few quick check calculations before finalizing a bid price for the site as shown below:

Site value as a percentage of the scheme's capital value	= 26%
Site value per hectare	= £1 175 000
Site value per acre	= £475000

Further calculations comparing ERV and building costs, and yields and building costs, etc. would provide an even more complete picture than that shown in Table 13.3, before a bid price for the site was finalized.

13.8 Capital profit, development yield and rent cover

Where a site is for sale at a stated price, developers often find it more useful to calculate the degree of profit realizable from development if the site were to be acquired at the asking price. In the same way, once a site is in a developer's ownership there is less relevance in undertaking a residual calculation to work out the development value of the site, as the site has been acquired and paid for; it is more useful to know the recoupment rent that would need to be charged in order for the scheme to break even or how much profit the developer is likely to make and whether this profit is reasonable or sufficient. This profit may be expressed in a variety of basic ways:

1. As a residual capital sum realizable if the completed scheme was sold;
2. As an annual sum, expressed as a yield, if the completed scheme was retained as an investment;
3. By the break even rent;
4. By the number of years it would take to eliminate the profit assuming a letting (and hence sale) were delayed.

This last calculation is known as rent cover, as it is applicable to those forward funding arrangements where the developer guarantees the rent from the end of the rent-void period (allowed in the appraisal) until the building is income producing.

Once the residual capital profit has been calculated it may be expressed as a percentage of the scheme's capital value, or its total development cost and then compared with the percentage allowances deemed appropriate according to the criteria laid down earlier in this chapter. Similarly, the annual sum, expressed as a yield, can be compared with the property's investment yield and the percentage difference calculated to see whether this provides an adequate mark up. Some examples will help to illustrate these alternative approaches.

The same information as contained in the appraisal shown earlier is used but it is assumed that the site can be acquired for £900 000 (including acquisition costs). The viability of the proposal can be calculated as follows:

Capital profit assuming a sale on completion

	£	£
Expected capital value		3 636 000
Expected total costs:		
Land cost (incl. acquisition costs)	900 000	
Building cost, fees, contingencies, etc.	1 663 000	
Finance @ 10% (compounded quarterly):		
land cost over development period		
of 1.75 years	170 000	
building costs, etc. over half building		
period plus full letting delay		
(i.e. 10.5 months in total)	150 000	
Letting and sale fees	133 000	
Total Costs	3 016 000	3 016 000
Residual capital profit		620 000

$$\text{Profit as a percentage of total costs} = \frac{620\ 000}{3\ 016\ 000} \times 100\% = \underline{20.56\%}$$

$$\text{Profit as a percentage of capital value} = \frac{620\ 000}{3\ 636\ 000} \times 100\% = \underline{17.05\%}$$

Not surprisingly the profit margins are higher than the minimum acceptable levels assumed previously, as the site cost is lower. However, where potentially high profit levels are likely some form of planning gain may be demanded by the local planning authority and there may be taxation implications (Capital Gains Tax).

Rent Cover

Residual capital profit	£620 000
Rental value	£300 000
Therefore rent cover	= 2.07 years

If the developer is guaranteeing the rent as part of a forward sale funding arrangement, as long as the building is let within two years a profit will still be realizable. In certain arrangements where the developer only guarantees the rent until the profit allowance is exhausted, the degree of rent cover is obviously vitally important from the funder's viewpoint.

As capital profit may be expressed as a percentage of capital value and as capital value is directly related to rent, it is clear that rent cover is merely a different way of looking at the same figures. It should be noted, however, that two schemes showing identical returns (when capital profit is expressed as a percentage of capital value) will have different degrees of rent cover if their investment yields are different. The higher the yield the lower the degree of rent cover; or, put another way, if the same rent cover is required a higher capital profit will be needed.

Break even rent

The break even rent is the rent that will ensure that no profit and no loss is made, i.e. where total capital value equals total costs. In this example, if the site cost is £900 000 (inclusive of fees), the break even rent is £62/m^2.

Annual profit assuming the scheme is retained as an investment

	£	£
Expected annual income		300 000
Expected total costs:		
Building costs, fees, etc. (as before)	1 663 000	
Land costs (incl. acquisition costs) (as above)	900 000	
Finance @ 10% p.a. (compounded quarterly)		
on land cost over development period		
of 1.75 years (as above)	170 000	
building costs etc. (as above)	150 000	
Letting and sale fees	133 000	
Total costs (as above)	3 016 000	

$$\text{Development yield} = \frac{300\,000}{3\,016\,000} \times 100\% = \underline{9.95\%} \text{ p.a.}$$

Just as the investment yield (8.25% p.a.) expresses income as a percentage of capital value (or cost to an investor of acquiring the scheme) so the development yield expresses income as a percentage of total costs incurred in creating the scheme. The difference between total costs and capital value is an expression of the developer's capital profit. Similarly, the difference between investment yield and development yield is an expression of the developer's annual profit. In this case it is 1.7% p.a. (9.95 − 8.25) which is a profit mark up of 20.6% (i.e. (1.7/8.25) × 100% = 20.6%) or the same as that shown above (after allowing for rounding errors).

In all the above forms of analysis where a higher rent than expected is achieved, resulting in a higher profit, and where the development is funded by a financial institution, it is probable that the funder would share in the increased profit realized often by allowing the developer only half the overage rent (i.e. the difference between the rent achieved and the rent initially expected).

13.9 Refinements to the residual valuation

The traditional residual method of evaluating development schemes is the method most commonly used in practice. This has been so for a long time despite its relative crudeness and the problems involved in its use as described in section 13.7. As property development has become more complex and computers and programmable calculators have become commonplace, more sophisticated techniques of appraisal and analysis are being used to refine, supplement or even supplant the traditional methods.

The main areas where the basic residual valuation can be made more sophisticated are:

1. The use of a cashflow apprach to measure more precisely the amount and timing of expenditure and income during development;
2. The use of sensitivity analysis to provide developers with a more detailed picture of a scheme's potential viability;
3. The use of forecasting techniques and probability analysis to assess the effect on viability of likely changes in the estimated values of key variables.

13.10 Cashflow approach

The cashflow approach is no more than a sophisticated, and potentially more accurate, refinement of the residual approach. A more detailed calculation of the scheme's total costs is achieved through a more accurate assessment of building (and hence of finance) costs in particular. However, the cashflow appraisal, whether a net present value (NPV), net terminal value (NTV), or period-by-period approach, also has the facility to extend

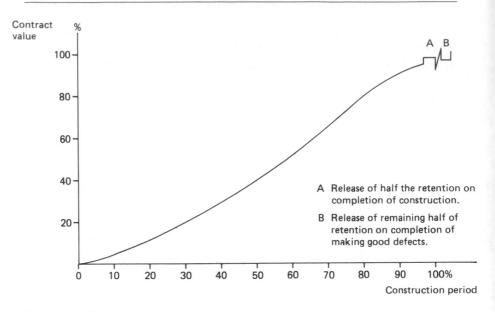

Figure 13.1 'S' curve showing typical construction cashflow.

the amount of information provided by appraising a scheme's viability, thus enabling decisions to be made with a greater degree of accuracy. The anticipated returns from property development can be compared more readily with alternative investments by means of their internal rates of return.

The basic difference between the simple residual and cashflow approaches is that in the latter method all development costs, and in particular building costs, are divided up into monthly, quarterly, half yearly, or yearly amounts, the net cashflows calculated and short-term finance allowed for separately in each period. In the simple residual approach the assumption was made that money borrowed at the start of development will incur interest over the whole period, whereas money borrowed towards the end of development will incur little or no interest, so that on average finance charges will be incurred on total costs over approximately half the period.

Owing to the incidence of costs during the building contract, it can be argued that a cashflow approach enables a more accurate appraisal to be undertaken even on relatively simple developments. The incidence of cost in most schemes conforms to an 'S' curve (Figure 13.1). It is usual at the start of building work for there to be a gradual build up of expenditure normally reaching a peak after 50–60% of the contract has elapsed with a tailing off towards the end. In a typical project, therefore, over half the building costs are incurred in the latter half of the project, and this will reduce interest accumulations as money will not need to be borrowed for so long.

A cashflow approach, however, is most useful in more complex schemes,

Table 13.4 Cashflow appraisal

Month	0	3	6	7	8	9	10	11	12	13	14	15	21	Total
Land cost incl. acquisition cost	900000													900000
Building cost incl. site layout, etc.				60000	90000	130000	170000	200000	220000	220000	200000	130000	20000	1440000
Professional fees		43000	43000	5800	5800	5800	5800	5800	5800	5800	5800	5800	5800	144000
Contingencies @ 5%		2100	2100	3300	4800	6800	8800	10300	11200	11200	10300	6800	1300	79000
Disposal fees													133000	133000
Total dev costs (Cashflow)	900000	45100	45100	69100	100600	142600	184600	216100	237000	237000	216100	142600	160100	2696000
Total income													+3636000	+3636000
Net cashflow	900000	45100	45100	69100	100600	142600	184600	216100	237000	237000	216100	142600	+3475900	
Interest charge @ 2.5% per quarter		22500	24200			27900			39300			57100	137000	308000
Total cashflow	900000	67600	69300	69100	100600	170500	184600	216100	276300	237000	216100	199700	+3338900	+632000
Cumulative flow	900000	967600	1036900	1106000	1206600	1377100	1561700	1777800	2054100	2291100	2507200	2706900	+632000	+632000

for example, phased developments such as large business parks and industrial estates where some buildings can be let or sold before others are
completed. The timing of these lettings and sales can be shown to be critical
to the profitability of the development. Grants, if payable at different stages
during development, can also be built into the appraisal, the total debt
outstanding at any point can be identified, and the total payback period and
date of peak cash outlay are easily shown, none of which are possible with
the traditional residual method.

The cashflow example (Table 13.4) illustrates the advantages of this
approach as applied to the development described in section 13.8, the
figures being modified so that building costs, professional fees, etc. are itemized on a monthly basis, from information obtained from analysis of
schemes of this size corresponding to the 'S' curve in Figure 13.1. It is
assumed that the site is acquired for £900 000 (inclusive of fees, etc.). In
Table 13.4 interest is compounded quarterly on the total net outlay incurred
in each period so that at any point during the development process the total
outstanding debt and the amount of interest owed is shown.

Interest is calculated with quarterly rests. For periods of less than three
months, interest is calculated as a proportion of the three-monthly figure –
i.e. it is calculated on a simple interest basis. for example, in month 9 the
figures are calculated as follows. The capital outstanding from previous
periods (cumulative flow) is the sum of £1 036 900 (the amount of money
owed after six months) plus £69 100 and £100 600 (the total development
costs in months 7 and 8) giving a total cumulative flow of £1 206 000.

Interest on £1 036 900 is calculated over three months,
    i.e. £1 036 900 × 0.025 = £25 900.
Interest on £69 100 is calculated over two months,
    i.e. £69 100 × 0.025 × 2/3 = £1 200.
Interest on £100 600 is calculated over one month,
    i.e. £100 600 × 0.025 × 1/3 = £800.
Total interest is therefore £27 900

The completed development, one and three quarter years after the start of
the scheme, is sold for £3 636 000. The cumulative cash outflow obtained
from Table 13.4 is £3 004 000 (£2 706 900 + £160 100 + £137 000), of which
£308 000 represents rolled-up interest. This leaves the developer with a
capital profit of £632 000 (compared to £620 000 in the earlier simple
residual example). This return represents 21% on costs, or 17.4% of capital
value, very similar to the figures assumed in the earlier residual valuation.
For a relatively straightforward scheme such as this, providing the monthly
expenditure conforms to average projects of this size and the site cost is a
significant proportion of total expenditure, the simple residual method can
be considered sufficiently accurate as the error in calculating finance charges

will be small. On more complex schemes and developments taking a number of years this is far less likely.

The measures of profitability, or return on costs as used above, are somewhat crude as they have to reflect a number of different factors, predominantly the degree of risk and the length of time before the profit (if any) is realized. So if one scheme takes four years to develop and another two years, a developer would obviously want a greater profit from the first scheme as compensation for the extra time before any money was earned. But how much extra profit?

An alternative (DCF) approach would be to determine what discount rate would equate total costs (excluding the profit allowance) to total capital value. This discount rate or internal rate of return (IRR) would therefore reflect the scheme's profitability relative to the actual incidence of costs and income and show the rate of return earned on funds invested in the project. This rate of return could then be compared with the rate of interest on borrowed money or on alternative investments; the minimum acceptable return would be the interest rate on borrowed money plus an acceptable margin for the degree of risk involved. The main advantage of this approach over the traditional approaches previously discussed is that the effect of time the developer has to wait before profit is received has been accounted for separately, so that the margin to be considered relates solely to the degree of risk involved. This approach also facilitates accurate comparison between different projects.

The following simplified example (Table 13.5) illustrates how the IRR is calculated: the approach is an iterative one – picking two trial discount rates on either side of the correct rate and interpolating between the two answers (net

Table 13.5 DCF analysis

Period (months)	Net cashflow	Discount factor @ 7%	@ 7.5%	Present value @ 7%	@ 7.5%
0	(900000)	1	1	(900000)	(900000)
3	(45100)	0.9346	0.9302	(42200)	(42000)
6	(45100)	0.8734	0.8653	(39400)	(39000)
9	(312300)	0.8163	0.805	(254900)	(251400)
12	(637700)	0.7629	0.7488	(486500)	(477500)
15	(595700)	0.713	0.6966	(424700)	(415000)
18	—	0.6663	0.648	—	—
21	(160100)	0.6227	0.6028	(99700)	(96500)
	3636000	0.6227	0.6028	2264100	2191800
	Net present value			16700	(29600)

present values) to work out the correct rate. Whilst this may look tedious the use of computers simplifies the calculation enormously.

The correct discount rate must therefore be between 7% and 7.5% and is obviously closer to 7% than 7.5%. An approximation of the correct figure is obtained as follows:

Difference between NPVs = £16 700 + £29 600 = £46 300

The 0.5% difference in discount rates which gives this difference of £46300 in NPVs can be apportioned to give the required discount rate. (A slight error occurs in this method of apportionment as the interest rates lie on a curve rather than a straight line as assumed here for ease of calculation. As long as the interest rates are not far apart any errors will be small and insignificant.)

$$\text{Either} \quad 7 + \frac{16700}{46300} \times 0.5 = 7.18\%$$

$$\text{or} \quad 7.5 - \frac{29600}{46300} \times 0.5 = 7.18\%$$

$$\text{Therefore true annual discount} = (1 + 0.0718)^4 - 1$$
$$= 32\%$$

If the cost of borrowed money was assumed to be 10% p.a. then an IRR of 32% p.a. shows a considerable margin to cover risk and profit. Most developers, who use this approach, would consider an IRR or risk adjusted discount rate of 20–25% to be acceptable, depending on the cost of borrowed money. The above analysis suggests therefore that this scheme is very profitable, mainly due to the low cost of acquiring the site.

There are many instances when a cashflow approach not only provides more information about the development but also provides a more accurate and detailed appraisal. Nevertheless, it would be a fallacy to say that a cash-flow approach is always suitable, or always more accurate, than a traditional residual. Whilst it is true that it can allow more accurately for the incidence of costs and returns, and therefore the actual interest incurred, the end result is only as accurate as the information used. If the timing of payments and recepts is uncertain, then an arbitrary or inaccurate assumption will result in an answer no more, and possibly less accurate, than that obtained by using a traditional residual. Where schemes are straightforward, where development periods are short, where interest rates charged on borrowed money are low and where appraisals are being undertaken at the early stages in the development process, possibly before a site has been purchased and an architect and quantity surveyor appointed, there will be little point in adopting a cashflow approach. But later, when consultants have been appointed and plans of the development are at a more advanced stage, more detailed and accurate information will be available, which may make a cashflow appraisal more applicable.

13.11 Sensitivity analysis

Because there are so many variables in any financial appraisal, great care must be taken as small changes in one or more of the variables can often exert a disproportionate effect on the residual answer. It is due to the inherent sensitivity of this method of valuation, and to reflect the general risks in undertaking a speculative development, that developers make an allowance of approximately 15–20% on top of estimated costs to provide a financial incentive. Most appraisals contain at least seven major variables, each of which must be estimated, and each of which must therefore be subject to change in the future.

Along with investment yield, rent is the most important variable, as small changes will exert the greatest change on profitability or residual site value. Accurately estimating rental levels is therefore crucial but difficult. It is difficult because it is unlikely that there will be sufficiently close direct comparables of recent lettings, even to act as a base estimate before any forecasting

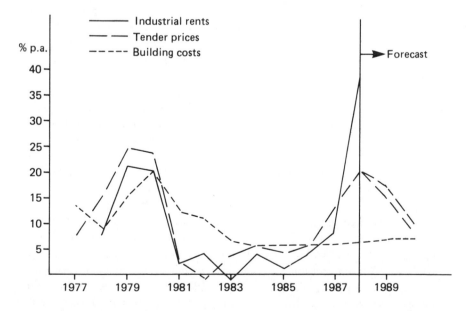

Figure 13.2 Change (% per annum) on previous reading for industrial rents, building costs and tender prices.
Sources: PRIME Healey & Baker; Davis Langdon & Everest.

is attempted of what might happen during the development period. For example, are the comparables of similar size, age, and condition? Are they in similar locations and do they possess the same level of finish and facilities? Was the rent achieved typical, or exceptional, or were there lengthy rent-free periods with the landlord paying fitting-out costs or were the leases of shorter than normal duration with break clauses? Obviously the experience and skill of the valuer will overcome some of these problems but, nevertheless, they will reduce the degree of certainty of any rental estimate.

Further problems arise because the development will not be available to let until some date in the future (probably one year minimum depending on the size of scheme) and so some estimate of what will happen during that period will have to be made even if it is not specifically made by projecting the rent to an expected future level. In the past rents have not increased in any uniform way but tend to move in cycles, and it may well be that a development started during an upturn in the cycle (boom) may be completed during the seemingly inevitable downturn (recession). Figure 13.2 illustrates this point very clearly, showing how the degree of rental growth has varied markedly from year to year over the last ten years. Furthermore, much more violent changes have occurred and are likely to continue to occur for particular properties in particular locations. Figure 13.2 also shows how construction costs can change dramatically over short periods of time, although they are probably less difficult to predict over the short term. The RICS (Building cost information service) and leading firms of quantity surveyors, for example, produce regular forecasts for two-year periods.

Small changes in investment yield levels can also have dramatic effects on residual calculations and yields can alter significantly during a two- or three-year development period.

The following simple analysis illustrates the degree of sensitivity of the development scheme referred to in section 13.7 and also how changes in certain key variables exert a greater effect on profitability than change in others.

It is very clear that rent, investment yield and, to a lesser extent, building costs are by far the most sensitive variables, with a 10% change in rent or investment yield, for example, altering profit by six times as much. The majority of development projects will show similar characteristics. In recognition of the key importance of these factors, developers seek to reduce their risk exposure, on occasion, by means of pre-lettings and/or forward sales coupled with fixed price building contracts.

The final column of Table 13.6 shows the scheme's downside risk where the development just breaks even. If rents fall by 15% then no profit will be made. In numerical terms this would mean rents of just over £63/m^2 rather than £75/m^2 (as estimated in the original appraisal). An investment yield increasing from 8.25% to 9.75% would have a similar effect.

The method of analysis illustrated above is basic but it does provide a

Table 13.6 Sensitivity analysis

Variable	Original value	New value (% change from original)	Return for risk and profit		Value of variable to achieve break even situation (% change from original)
			Amount (original figure £545 000)	% of total costs (original figure 17.65%)	
Rent	£75/m²	£67.5/m² (−10%)	£195 000	6.3%	£63.5/m² (15%)
Yield	8.25%	9.07% (+10%)	£223 000	7.2%	9.75% (18.2%)
Building and site layout costs	£360/m²	£396/m² (+10%)	£364 000	11.1%	£468/m² (30%)
Finance rate	10%	12% (+20%)	£475 000	15%	
Building period	9 months	12 months (+33%)	£494 000	15.7%	
Letting period	6 months	9 months (+50%)	£472 000	14.9%	
Pre-building period	6 months	9 months (+33%)	£517 000	16.6%	
Land	£940 000	£1 034 000 (+10%)	£430 000	13.4%	

developer with additional useful information about a scheme's viability and sensitivity to changes in estimates. It highlights the key variables, the degree of sensitivity and the extreme conditions which have to occur before profit margin is completely eroded. What it fails to consider is the rather more likely occurrence of combinations of a number of variables changing simultaneously rather than in isolation and secondly, the probability of these changes occurring.

13.12 Scenarios and forecasting

A simple approach, used by some developers, would be to examine the effect of various permutations:

1. tomorrow's cost and tomorrow's rent;
2. tomorrow's cost and today's rent (safe or pessimistic view – pessimistic scenario (a) in Table 13.7);

3.   today's cost and tomorrow's rent (optimistic view – optimistic scenario (b) in Table 13.7);

More detailed scenarios involving predicted values for each variable could also be undertaken. In practice, professional judgement will be crucial in selecting reasonable estimates (forecasting) based on expert advice, good records and knowledge of the economy and the local property market. These calculations (Tables 13.7 and 13.8) would show the developer various outcomes that his expertise suggests are probable, thus providing a more complete picture of potential viability, and should enable a more informed decision to be made about proceeding with the scheme in its present form at the present time.

Many developers are sceptical about this type of approach as it involves predicting values for rents, yields and costs, etc. and they question the validity and usefulness of the results, which obviously are so highly dependent on the accuracy of the inputs, This is understandable, particularly in view of the experiences of many developers in the early/mid-1970s when bullish views of the property market, involving optimistic rental projections, caused the downfall of many property companies. Nevertheless it is one thing to buy a site based on a residual valuation using predicted, rather than prevailing values, but quite another to analyse a possible site purchase and appraise financial viability by examining the effect of different forecasted values occurring. Whilst it is certainly true that predictions are difficult to make accurately and have in the past sometimes been notoriously inaccurate, more research into the property market, rental growth trends and the movement of tender prices and building costs, is now undertaken

Table 13.7 Scenarios

Variable	Optimistic scenario		Realistic scenario	Original estimate	Pessimistic scenario	
	(a)	(b)			(a)	(b)
Rental growth*	7% pa	5% pa	5% pa	–	–	4% pa
Investment yield	8%	8.25%	8.25%	8.25%	8.25%	8.5% pa
Tender/building costs increase (*)	7%	–	8% pa	–	8% pa	9% pa
Finance rate	9%	10%	10%	10%	10%	12%
Building period	9 mths	9 mths	9 mths	9 mths	9 mths	12 mths
Letting period	3 mths	6 mths	6 mths	6 mths	6 mths	9 mths
Prebuilding period	3 mths	6 mths	6 mths	6 mths	6 mths	6 mths

*to completion of building period
(*) to mid-point of building period (average allowance)

by the larger estate agency and quantity surveying firms, financial institutions and the RICS, research which a decade or more ago was largely non-existent. Much more is now known about the property market and the uncertainties which pervade predictions of rental growth and cost inflation have been reduced.

The scenarios in Table 13.7 (again applied to the same development scheme) have been chosen to illustrate the effect of combining predicted values for different variables. These illustrations are not intended to represent complete extremes of circumstances which the developer could face, but situations which could reasonably occur.

The results of this analysis (Table 13.8) show the developer, in some detail, how sensitive this scheme is, as relatively small changes in each variable when combined together lead to dramatically altered profit margins. It is also noteworthy that in the realistic, and therefore most expected scenario, even though a safe view of rental growth, *vis-à-vis* cost inflation, has been taken (possibly even a pessimistic view) profit margins are improved when compared to the original appraisal. The main reason for this is that most developments tend to be let and sold *at the end* of the development period, therefore benefiting from growth in values throughout the whole period, whereas building costs are incurred and paid at stages *during* the development period and land/site costs are paid as a fixed cost before building commences. In many cases, where letting occurs when the scheme is completed, annual rental growth considerably less than that of annual cost inflation will still give a similar residual profit as shown by the traditional approach of using present day values and costs.

Further extensions and refinements of this approach can be adopted, by incorporating realistic, optimistic and pessimistic values for *each* variable and then combining these values to give a wider range of possible results. Similarly, the probability of these estimates occurring or the standard deviation for each variable can be assessed, so enabling more complex and

Table 13.8

	Optimistic scenario		Realistic scenario	Original estimate	Pessimistic scenario	
	(a)	(b)			(a)	(b)
Developer's return	960 000	779 000	653 000	545 000	420 000	246 000
% of CV	23.9%	20.1%	16.8%	15%	11.5%	6.6%
% of total costs	31.5%	25.1%	20.25%	17.65%	13%	7.05%
Increase over original estimate	76%	43%	20%	—	−23%	−55%

refined models to be developed, including simulation exercises, to show the developer the probability of various levels of profit occurring. Whilst advanced statistical techniques exist which, with the aid of computers, make such analysis relatively straightforward, one should not lose sight of the truism that an answer is only as reliable as the accuracy of the inputs. Nevertheless, off-the-shelf computer software is now readily available which will undertake Monte Carlo simulations, for example, and these programs are now used by many leading developers. See Morley (1988) or Byrne and Cadman (1984) for a more detailed examination of these techniques.)

13.13 Conclusion

Recent years have seen considerable change in the way property development is undertaken, the degree of market research required and the methods used to appraise viability, not to mention changes to developers themselves. Major structural and technological change will necessitate even greater reliance on improved methods and techniques, forcing industrial and business space development to become a more sophisticated process. Undoubtedly, hunch, intuition and flair will always remain an essential ingredient of the successful developer, but these qualities are reinforced and backed up by techniques which are now increasingly commonplace in the general investment market and which have been used in other areas of the economy for decades.

References

See end of Chapter 14.

14

Public sector land ownership

14.1 Introduction

Chapter 11 discussed in some detail the various options available to owners of land wishing to implement development, ranging from outright freehold or long leasehold disposal, to various types of partnership arrangements, and even direct development where freehold land ownership was retained. Chapter 13 examined financial appraisal methods appropriate to outright disposal; this chapter is concerned with the different financial arrangements available in partnership agreements and direct development by local authorities. As with any financial arrangement there are an almost infinite variety of alternatives but the appraisals discussed in this chapter will relate to the main types of arrangement, with a brief examination of what is involved and the implications to the various parties.

The chapter is subdivided into three parts. The first (sections 14.2–14.5) examines methods of appraisal relevant to schemes which are inherently attractive to the private sector, for example, industrial and business park schemes in the southern half of the country particularly with the majority of unit sizes in excess of $500 \, m^2$. Traditional four slice or side by side partnership arrangements would be applicable. The second part (sections 14.6–14.8) covers the implementation of schemes which the private sector consider to be more risky. Typically these might be small unit developments within urban areas in the more prosperous parts of the country. Location, tenants' covenants, management problems, etc. will necessitate greater public sector involvement to reduce development and investment risks to the private sector. The most common type of partnership agreement would be where the local authority take a leaseback of the completed development, subletting the individual units themselves. The third part (section 14.9) covers schemes of all types in areas of the country where the private sector is reluctant to venture because of the much greater development risk, the low level of rents and the lack of viability. In these areas direct development by local authorities and other public sector agencies is often unavoidable.

14.2 Four slice and side by side partnership agreements – a comparison

Early partnership agreements in the immediate post-war period were usually simple arrangements with 99-year leases at a fixed rent or with infrequent rent reviews. The word partnership was even more of a misnomer than it often is today. From the late 1960s onwards, however, more sophisticated arrangements became commonplace in town centre redevelopment and these four slice arrangements were subsequently used for industrial and office developments and are still commonplace today. The term 'four slice' is used to illustrate the priority attached to the apportionment of different tranches of income from a completed development. The first tranche or slice (the ground rent) is payable to the freeholder. Tranches two and three covering development costs and development profit are payable to the developer. The fourth tranche is any excess profit on completion of development which is apportioned between the freeholder and developer. This is illustrated and compared with a side by side arrangement in Figure 14.1. Increasingly the side by side partnership arrangements are favoured by developers where income and risk from developments are shared in some predetermined ratio sometimes without the guarantee of any minimum income or initial ground rent to the freehold landowner (usually the local authority). Side by side arrangements are more of a true partnership and are therefore favoured by the private sector as risks are shared more equally. As a result the development return required is usually lower providing the freeholder with a potentially greater income, which may be attractive to some freeholders even though the risks are greater. A more detailed explanation and comparison between these two types of partnership arrangement occurs later in section 14.4, but first the basic principles must be examined.

The basic differences between an appraisal to calculate ground rent rather than capital site value are that first, the calculations are undertaken on an annual rather than a capital basis and second, from the developer's point of view, capital costs are obviously reduced as the site is not acquired, no capital is at risk and no finance charges are incurred. Although a rent (annual cost) would be paid for the site this is normally not payable until completion of development when the scheme should be revenue producing. For these reasons in certain locations some developers prefer such an arrangement to outright freehold ownership, particularly as the landowner becomes an investor in the scheme. However, it is probably true to say that many developers would prefer freehold ownership due to the greater control over development and subsequent investment and this factor will be reflected in the return they will require.

A typical appraisal, using the same scheme as in Chapter 13 but assuming a 125-year lease with say five-year rent reviews, might be as follows:

Ground Rent Calculation
  Estimated rental value
    expected rents from occupational tenants   £300 000 p.a.
    (Chapter 13)
  Estimated development costs
    building costs, fees, finance, etc.,   £1 946 000
    but excl. land (Chapter 13)
  Development yield
    initial leasehold investment yield or long-term
    finance rate of say 9% plus annual return for
    risk and profit of say 2% p.a. (ignore annual
    sinking fund to recoup capital as insignificant
    over 125 years lease and the leasehold yield is
    adjusted accordingly). Therefore total deve-
    lopment yield required is 11.0% p.a.   0.11   £214 000 p.a.

  Ground rent (commencing on completion
    of development)   £86 000 p.a.

Income apportionment, or gearing, at each rent review
(proportionate sharing):

Developer: $\dfrac{214\,000}{300\,000} \times 100\% = 71.3\%$

Freeholder: $\dfrac{86\,000}{300\,000} \times 100\% = 28.7\%$

If either the freehold landowner, or the developer, wished to sell their interests, *on completion of the scheme,* then the following valuations would apply:

Freeholder

Ground rent income	£86 000 p.a.
YP in perp. @ say 8%	12.5
Capital value	£1 075 000

Developer

Full rental value	£300 000 p.a.
less ground rent	£86 000 p.a.
Net income	£214 000 p.a.

YP in perp. @ say 9%	11.11
Capital value	£2378000
Total costs (as above)	£1946000
Therefore profit	£432000

Therefore Return on Costs $\dfrac{£432000}{£1946000} \times 100\% = 22.2\%$

A useful additional calculation for the freeholder is to compare the capital value of the land after development (shown above) with the site's book value or cost of acquisition. Where the site has been acquired to initiate development it is useful to compare the initial return from land acquisition as follows:

Site purchase cost	say	£700000
Cost of infrastructure, etc.	say	£100000
Interest on costs until completion of development (and receipt of ground rent)	say	£250000
Total cost		£1050000
Ground rent or income		£86000 p.a.

Initial return to freeholder $\dfrac{£86000}{£1050000} \times 100\% = 8.2\%$

An alternative approach would be to compare annual interest payments, incurred by purchasing the site, with annual income (ground rent).

Total site costs on completion of development	£1050000
Annual interest repayments @ say 10% p.a.	0.10
	£105000 p.a.
Initial ground rent	£86000 p.a.
Therefore initial deficit	£19000 p.a.

Although an initial deficit occurs in this example, the interest repayments are fixed so any growth in rack rental value (and hence ground rental value) will reduce this deficit and hopefully produce a surplus at subsequent rent reviews. But it should be emphasized that criteria other than financial ones may be important to the freeholder, if it is a local authority, in assessing a scheme. Even if a poor financial return appears probable, the local authority may well proceed with implementation in order to obtain other benefits (e.g. employment). The scheme might act as a catalyst and stimulate other schemes to proceed which otherwise would not have occurred.

14.3 Explanation of the ground rent calculation

Estimated costs

£1 946 000 is the total cost excluding an allowance to cover developer's risk and profit (allowed for separately in the development yield) and excluding the cost of site acquisition, as obviously this cost will not now be incurred by the developer. The cost of fees for selling the completed investment has been included in the figure of total costs. As explained previously, the inclusion of this item of expenditure will depend on the developer's intention. But if the developer does not intend to sell the completed investment then funding costs will be incurred. If the developer is a financial institution then again no sale will occur as the scheme will be retained as a long-term investment. However, in this situation a project management fee might be included.

By not acquiring the freehold interest in the site a developer will save not only capital outlay, but also associated legal fees, agents' fees, stamp duty and finance charges. However, legal fees, agents' fees and stamp duty will still be incurred on the preparation of the building agreement and acquisition of the ground lease, although these have not been shown separately in the above appraisal.

Development yield

The total required development yield must be calculated to ensure that the developer receives sufficient return before calculating the residual ground rent. There are two components to this yield – the investment yield and the developer's return for risk and profit.

In the original appraisal in Chapter 13, an investment yield of 8.25% was used to capitalize rack rental income where the freehold investment value of the completed development was required. In this example the developer will have a less attractive interest, albeit a very long leasehold interest of 125 years. Nevertheless, leasehold investments are generally considered less attractive by investors as they do not have total control of the investment, there is more management, the investment is a wasting asset, and rent payable may be reviewable on an upward only basis. Yields tend to be slightly higher, depending on the length of lease, the convenants in the lease and the type of property. For example, if the rent is reviewable on an upward only basis and the gearing relates to rental value rather than rental income and the gearing is substantial – say over 10–15% of the rack rental value – then the investment yield could be substantially above that for a freehold. An extra 0.75% has been allowed here. Depending on the state of the property market and the availability of investments, together with the points made above, this margin could vary.

In Chapter 13 it was assumed that a developer of this scheme would want a margin to cover risk and profit of about 17.5% on costs. Although this margin will vary according to the particular scheme, the state of the market, the degree of pre-lets, or forward funding, and between different developers, 17.5% was taken as a norm for this type and size of scheme. With an 8.25% freehold investment yield the developer would therefore require an overall development yield of about 9.7% (i.e. 8.25% × 1.175). In the ground rent appraisal shown above, a development yield of 11% was assumed which shows the developer a higher 22% mark up on the leasehold investment yield of 9% (i.e. 9% × 1.22 = 11%)) for reasons explained below.

It could be argued that, as this scheme is being undertaken in partnership with the freeholder and the developer's outlay is reduced by the savings in land cost, the developer might be prepared to accept a slightly lower profit margin but much will depend on the arrangements regarding rental guarantees, subsequent gearing, sharing of excess profits and the basis of rent review as discussed more fully below. While it is true that the developer's *capital* outlay is reduced by the saving in land cost, an additional *annual* expenditure (ground rent) will be incurred. Where the ground rent is a guaranteed minimum figure and excess profits are to be shared at the participation date (shortly after completion of development, i.e. a four slice arrangement) a developer should, therefore, require a higher percentage profit margin on costs to compensate.

The development yield required is made up of the investment yield and the annual return to cover risk and profit. In this example, so long as the developer achieves an 11% yield he will be prepared to give away the remainder of annual income as a ground rent payment to the landowner (e.g. the local authority). Obviously the developer would like to achieve a higher yield than 11% but in a competitive tender situation this is the minimum figure needed to make the scheme worthwhile. It is arguable that if the developer were not in a competitive situation then a negotiated settlement might result in a higher development yield.

Ground rent and equity sharing (or gearing)

The ground rent would normally become payable after completion of the development or after a specified time limit equivalent to the estimated length of the development period. During the development period a pepper-corn (or zero) rent would be paid, so that the ground rent effectively becomes payable when income is received by the developer from letting the completed scheme. Traditionally, in many partnership agreements, the free-holders'/local authority's interest would be safeguarded by ensuring that the estimated ground rent was a minimum figure which could be increased if and when the rental income from the occupational tenants increased, but could not be reduced if the scheme was less profitable than the developer expected, i.e. the freeholder's income was the first charge or slice out of total income from the scheme (Figure 14.2).

If the ground rent or gearing ratio was not a minimum figure there would be a danger that the developer would overbid initially to win the nomination and then reduce the ground rent subsequently when actual costs and rents were known. Nevertheless, in certain situations where tenant demand is less strong and the development risks are perceived to be greater, developers may be reluctant to take the risk of offering a minimum ground rent. In this situation the freeholder may be faced with either having to abandon the scheme or else entering into a true side by side partnership, sharing the development risk with the developer. Even where risks are not great developers may still make offers on a side by side basis in return for a more attractive gearing ratio to the freeholder.

Equity sharing or gearing effectively means that the initial relationship between ground rent and rack rent is maintained at every rent review. Every time the rack rent increases the ground rent would also increase by the same percentage amount. It is important to clarify the term 'rack rent' in this context as there is an important difference between rental value and rental income. From a freeholder's viewpoint the former is desirable as it ensures that ground rents are calculated as a percentage of the maximum rental income that the scheme could produce if there were no voids which could be caused by a down turn in the market, natural turnover of tenants or refur-bishment. A developer may try and resist the use of the term rental value, particularly if it is coupled with a high gearing ratio, arguing that a higher development yield (and hence lower ground rent) would result. The reasons for this are illustrated in Figure 14.1 which helps to explain why side by side agreements, where *actual* income is apportioned are growing in popularity.

Although the equity sharing arrangement in the above example is a common and simple method of apportioning future rental growth it is by no means the only method. In some cases a local authority may prefer to receive a lower initial ground rent in return for a higher share of future rental growth. A developer may also prefer this arrangement in certain situ-ations particularly if the ground rent is to be a minimum figure as the developer will then be guaranteeing a smaller sum (less initial risk), but giving away more equity if the scheme is successful.

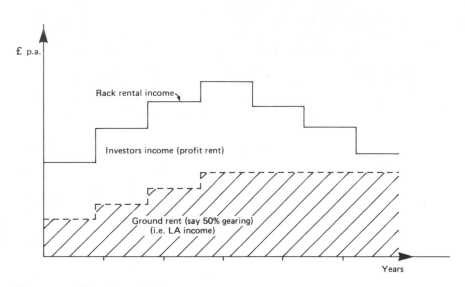

Figure 14.1 The effect of high gearing and voids.

14.4 Participation clauses and four slice agreements

Four slice agreements contain a participation clause which involves the residual calculation being reworked shortly after completion of development, using actual rather than estimated figures enabling the freeholder/local authority to participate in any excess profits. Due to possible inflation of costs and rents during the development period it is likely that actual rents and possibly costs will differ from those estimated. The purpose of the participation clause is to reflect these changes in a revised ground rent and possibly a revised equity share. This is particularly likely in larger and more complex schemes due to the length of time that often elapses between the tender date, when the ground rent offer is made, and completion date when the scheme is income producing. The revised equity share will then determine the local authority's income at subsequent rent reviews.

If the initial agreement stated that the estimated ground rent was to be a minimum figure (subject to upward only review and therefore a first charge out of rental income) then the object of the participation clause will be to increase the ground rent once the developer has achieved the stated return required on actual development cost (yield protection). If costs have increased to such an extent that the developer is unable to achieve this return then the initially agreed ground rent will remain as a minimum figure and any reduction in income will be suffered by the developer as this is part of the risk of development. If the scheme is not fully let at participation, then many partnership agreements allow for rental value to be used for the

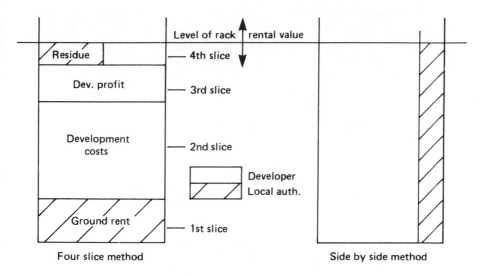

Figure 14.2 Four slice and side by side partnership agreements.

unlet parts, in order to obtain the total rent as mentioned earlier.

The differences between a four slice agreement and a side by side agreement now become very clear and are illustrated in Figure 14.2 and the numerical examples which follow. With a side by side agreement the calculation of ground rent on completion of development is more straightforward. The gearing ratio or equity share is simply applied to the actual income paid by the occupational tenants to give the ground rent, although in some agreements this simplicity is complicated by the inclusion of a minimum ground rent to give some additional security to the local authority.

The four slice agreement discussed and illustrated in Figure 14.2 enables both partners to participate in excess profits generated by successful schemes. Occasionally more complex sharing arrangements have been used in practice where the nature of the scheme and its location warrant them. For example, the 50:50 split of the fourth slice, or excess profit, may be subject to cut offs if the developer's return exceeds certain thresholds. Above such thresholds the freeholder may be entitled to a greater share than 50%. Such clauses are particularly relevant in the event of the developer selling the completed scheme.

The following numerical examples illustrate the calculation of a revised ground rent comparing a standard four slice with a side by side agreement, using the appraisal at the start of this chapter as the basis for the initial ground rent offer. In reality it is probable that some developers will be prepared to accept different development yields depending on whether their offers were on a four slice or a side by side basis for the reasons mentioned earlier. This difference in development yield could be as much as half of one per cent or more. For the purpose of comparison – in the following examples – this has been ignored and the original appraisal figures are assumed to be applicable for both offers. In the four slice agreement it is assumed that the initial agreement (contained in the development brief) stated that at the participation date any equity would be apportioned 50:50 (the usual apportionment).

The first example assumes that from the date the tender was made rental values have increased by say 20% and development costs by say 10%.

		Four slice	Side by Side
Actual income (+20%)		£360000 p.a.	£360000 p.a.
Less initial ground rent		£86000 p.a.	
Net income		£274000 p.a.	
Actual costs (+10%)	£2141000		
Development yield @ 11%	0.11	£235500 p.a.	
Residue or excess		£38500 p.a.	
50% of excess to freeholder (i.e. say £19000)			
Therefore revised ground rent		£105000 p.a.	£103000 p.a.

$$\text{Therefore revised equity share to freeholder} = \frac{105000}{360000} \times 100\%$$

$$= 29.2\%$$
(28.7% originally)

(28.7% gearing)

Where the scheme is more profitable than expected (and assuming no difference in development yield) the freeholder benefits more from a four slice agreement. If rental value differed from rental income (due to part of the scheme being slow to let) then the difference between the two agreements would be more marked. However, where costs and rents increase by the same percentage amount, the local authority could benefit more from a side by side agreement unless the four slice agreement stated that the gearing ratio could not be reduced (see below).

	Four slice	Side by Side
Actual income (+10%)	£330000 p.a.	£330000 p.a.
Less initial ground rent	£86000 p.a.	
Net income	£244000 p.a.	

Actual costs (+10%)	£2141000		
Development yield @ 11%	0.11	£235500 p.a.	
Residue or excess		£8500 p.a.	
50% of excess to freeholder			
(i.e. say £4000)			
Therefore revised ground rent		£90000 p.a.	£95000 p.a.

$$\text{Therefore revised equity share to freeholder} = \frac{90000}{330000} \times 100\%$$

$$= 27.3\%$$
(28.7% initially)

A problem arises here for the developer, with the four slice agreement, because the revised ground rent is higher than the previously agreed minimum figure, whereas the revised equity share is lower. Many partnership agreements state that the original equity share would apply for the remainder of the lease (even though a higher ground rent is also paid) as it was stated to be a minimum figure. As the difference in this example is small the burden to the developer is not that great but in the third example below serious problems could result and some developers will resist the possibility of a minimum equity sharing arrangement.

		Four slice	Side by side
Actual income (−5% due to voids			
Rental value remains unaltered)		£300000 p.a.	£285000 p.a.
Less ground rent		£86000 p.a.	
Therefore net income		£214000 p.a.	
Actual costs (+10%)	£2141000		
Development yield @ 11%	0.11	£235000 p.a.	
Residue		− £21500 p.a.	
Therefore revised ground rent		£86000 p.a.	£82000 p.a.
		(no change)	

In this example there is no excess to share and the initial ground rent and equity share remain unaltered in the four slice arrangement if the original agreement stated them to be minimum figures, but in the side by side arrangement the ground rent actually decreases as rental income has decreased. The developer in these situations could therefore be faced with a reduced profit:

	Four slice	Side by side
Rental income	£285000 p.a.	£285000 p.a.
less revised ground rent	£86000 p.a.	£82000 p.a.
Developer's income	£199000 p.a.	£203000 p.a.

Therefore actual development

yield
$$= \frac{199000}{2141000} \times 100\%$$
$$= \frac{203000}{2141000} \times 100\%$$

$$= 9.3\%$$
$$= 9.5\%$$

Assuming an investment yield of 9% (as before) this results in an annual return to the developer of 0.3% (a profit mark up on outlay of only 3.3%) in the four slice arrangement, and in the side by side arrangement a profit mark up of 5.5%, assuming no minimum ground rent had been guaranteed.

In extreme situations where there is an even wider margin between changes in costs and rents this return could be eliminated altogether, possibly even resulting in a loss. One could argue that this is all part of the risk involved in development which the developer was prepared to accept and has to accept in all speculative development situations; that is why typically a 15–20% mark up is allowed in case this situation arises. However, whilst a large development company or institutional developer could bear a loss due to financial reserves and profits from other developments, small companies might not be able to and could go bankrupt, possibly leaving part of the development unfinished. In such situations the freeholder will be under considerable pressure to allow a renegotiation of terms more favourable to the developer. Therefore, when competing bids are analysed considerable attention must be paid not just to the ground rent offered, but the component parts of the offer, their reasonableness and probable future scenarios to assess the local authority's exposure. (This is discussed in more detail in Chapter 11.)

The necessity to undertake more than a skin deep analysis is clearly demonstrated in the bids from two developers shown below. Both offers are on a four slice basis, Developer A's is based on current levels of costs and rents (the appraisal used previously) and Developer B's is based on projected costs and rents. Although Developer B is looking for a higher return (development yield of 12% rather than 11%) his ground rent offer is higher and superficially more attractive.

If Developer B's estimates of future increases in costs and rental value prove correct the offer of ground rent and equity share will remain unaltered at the participation date but Developer A's offer will be revised to the

	Developer A		Developer B	
	£	£ p.a.	£	£ p.a.
Estimated rental value		300 000		360 000 (+20%)
Estimated development costs	1 946 000		2 141 000 (+10%)	
Development yield	11%	214 000	12%	257 000
Ground rent		86 000		103 000
Freeholder equity	$\dfrac{86\,000}{300\,000}=28.7\%$		$\dfrac{103\,000}{360\,000}=28.6\%$	

figures shown previously (where the four slice and side by side arrangements were compared). The revised ground rent would be £105 000 and the new equity share, to be used at every subsequent rent review, would be 29.2%. If this scenario were to occur the local authority would actually be better off with the offer from Developer A although a comparison of the initial offers suggested Developer B's offer was better. Of course, if an outcome less favourable than Developer B's forecast actually occurred, the guaranteed minimum offer from Developer B would stand and the local authority could be better off accepting that offer, although in that situation the developer's profits would be reduced and renegotiation of terms might be attempted. This emphasizes the importance of the developer's financial backing and reliability.

14.5 Ground rent and premiums

In many partnership schemes local authorities prefer to receive a premium instead of an annual income or a smaller premium in return for accepting a lower ground rent. This capital sum can then be used to supplement capital spending elsewhere within their area or possibly to help purchase the site itself. In an era of public sector expenditure and borrowing contraints this desire may be particularly strong. A capital premium could be looked upon as being a non site-specific planning gain. One problem with this approach is that under the 1980 Local Government, Planning and Land Act (and subsequent circulars) only 30% of this capital receipt can be used to supplement local authority capital spending in the same financial year (see Chapter 12 for a detailed discussion of this point and changes proposed by central government). Alternatively some element of planning gain could be incorporated within the development scheme or on a different site similarly increasing the developer's costs and so reducing the ground rent payable, but possibly without the local authority suffering from the 30% rule.

Another reason behind the use of premiums is to keep the gearing ratio between ground rent and rack rental value low. Where site value forms a large percentage of the total project's value (e.g. in high value locations) then funding institutions will resist an unfavourable gearing where the freeholder receives more than about 10–15% of rack rental *value* unless the lease is not subject to upward only rent reviews (i.e. on a side by side basis where actual *income* is shared). If a large proportion of the equity was receivable by the freeholder, the leaseholder could be left with an exposed top slice income in the event of a subsequent decline in rack rental income (i.e. if rents fall or there are voids) and where the ground rent is reviewable on an upwards only basis. This point was illustrated in Figure 14.1. A premium reduces gearing making the basis of rent reviews less important and consequently may lead to a reduction in development yield required. (A 0.5% reduction has been assumed in the calculation below.)

Another solution to the problem, as mentioned above, is to have true side by side sharing of rack rental *income* which also may lead to a substantial reduction in development yield required.

Revised ground rent calculation assuming a premium of £750 000 payable at the commencement of development

Estimated income (as before)		£300 000 p.a.
Estimated costs		
development costs (as before)	£1 946 000	
premium	£750 000	
finance on premium over 1.25 years @ 10% p.a. compounded quarterly	£99 000	
Total costs	£2 795 000	
Development yield @ say 10.5%	0.105	£293 500 p.a.
Revised ground rent		£6 500 p.a.

In this situation the apportionment of future rental growth might be:

Freeholder
$$\frac{6500}{300000} \times 100\% = 2.2\%$$

Developer
$$\frac{293500}{300000} \times 100\% = 97.8\%$$

It is clear that in return for receiving a premium, the initial ground rent and subsequent equity share to the freeholder is reduced. Had the premium been payable on completion of development rather than at its commencement the developer would not have to allow for the cost of short-term finance, overall development costs would therefore be reduced and either a higher premium or larger ground rent would result.

In certain circumstances, depending on the relationship between interest rates and development yields, it might be more beneficial for a local authority to maximize its ground rent and equity share and, rather than accept a premium from the developer, borrow the money elsewhere at current money market interest rates, or sell other assets which it may own to realize the capital. However, due to the tight controls on capital expenditure and the fact that only 30% of capital receipts could be used in this way in the year in which they were realized (again due to central government constraints on capital spendng), this option is perhaps currently rather theoretical. (See Chapter 12 for changes in these regulations.)

14.6 Two slice and lease/leaseback agreements

In some partnership schemes in areas where development risks are considered by the private sector to be unacceptably high a variety of alternative arrangements of income and risk sharing have evolved. The side by side basis mentioned in the previous section is an example of this because the developer/investor is not necesarily guaranteeing the freeholder a minimum ground rent on completion of development and the freeholder is sharing in actual income received from occupational tenants, which means that its income could fall as well as rise throughout the period of the ground lease.

Another alternative to the traditional four slice arrangement – often referred to as a three slice arrangement – is for the developer to guarantee a minimum ground rent, but a lower figure (thus reducing his risks), take development costs as a second slice or charge and then apportion the resultant profit (third slice) in some predetermined side by side basis with the freeholder. As a lower ground rent is guaranteed, the developer is more likely to achieve a profit if the scheme is not particularly successful, and in this situation, a lower gearing ratio would apply to all subsequent income received from occupational tenants of the completed development. Other permutations of the three and four slice arrangement exist which all have the same objective, namely to reduce the developer's guarantees and make it more likely that a return on costs will be achieved.

The other side of the coin obviously is that if the scheme is more successful than expected it is the freeholder which benefits more than it would have done under the traditional four slice arrangement.

14.7 Two slice arrangements

Where the local authorities are considering the implementation of small unit industrial schemes in locations where demand is less certain, other arrangements will need to be considered where even more of the development risk is shouldered by the local authority. This may be achieved by the local authority guaranteeing the rack rental income, but without taking a pre-let of the units, or by agreeing to take an overriding lease of the scheme and then being responsible for subletting to individual tenants, or by not requiring a ground rent as a first charge out of income and in effect guaranteeing the developer a return on costs before the local authority receives any income from their freehold land ownership. These latter two alternatives, which are in many ways similar, are illustrated in Figure 14.3 and clearly show how the local authority now has a much more risky top slice income, whereas the developer has a much more secure bottom slice income.

In a straightforward two slice arrangement the ground rent, rather than being a predetermined guaranteed minimum figure, would typically be calculated on completion of development, with the developer receiving a predetermined return on actual costs incurred, but probably subject to maximum and minimum gearing to avoid the developer having an absolute guarantee of profit and to avoid the local authority receiving 100% of any excess profit. So, in practice, this approach (not widely used) would be rather more complex than Figure 14.2 suggests.

For small unit industrial developments even the above two slice agreement has problems, which are more related to the subsequent investment rather than the development process itself. This is obviously important to a developer because, on completion of the development, the scheme will either be sold on to an investor or refinanced enabling the developer to become a long-term investor. The problem concerns potential voids, tenants

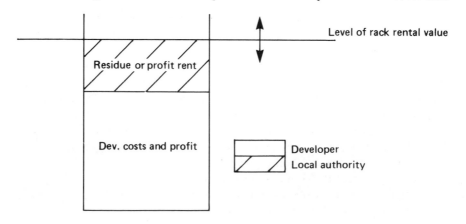

Figure 14.3 Two slice and lease/leaseback arrangement.

covenants and the management of an investment with numerous tenants. Small unit and work space schemes by definition will be let to small companies, many of whom may well be start-ups with no track record. Numerous tenants with poor covenants is not an appealing investment recipe, particularly to institutional investors, who like a limited number of tenants all of high calibre and long leases (25 years typically), with rent reviews on an upwards only basis to give good income security and minimal management. Small companies will not be prepared to take long leases. They want short leases, ideally with 'easy in, easy out' terms. Where such leases are available, not surprisingly there will be a high turnover of tenants and potential income voids until new tenants are found.

14.8 Lease and leaseback agreements

The solution to these problems is for the local authority to take a leaseback of the entire scheme and itself sublet to occupational tenants. This reduces the development risk as the developer has a pre-let at a known rent to one tenant of high calibre. It also makes a much more saleable investment and makes funding easier. Whilst the local authority is now taking a substantial risk it does at least gain by having control over letting, in terms of choosing the tenants it wants and on the lease terms it wants. It should receive a profit rent (difference between the rent received from occupational tenants and the rent paid to the developer) admittedly as a top slice income rather than as a secure bottom slice income, but at least it does not have the inconvenience of undertaking the actual development (which it would do if it were developing itself).

Alternatively, to circumvent the controls on capital expenditure and borrowing (see Chapter 12 for the regulations concerning leasebacks) the local authority might enter into a management agreement on behalf of the developer, charging a fee which is equivalent to the profit rent on a leaseback basis. The local authority would still control letting but would only have to take an actual lease if it defaulted on the management agreement.

A typical leaseback arrangement is shown in the numerical example below, based (for ease of comparison) on the same scheme as used at the start of this chapter. If the scheme was built as small units (i.e. below $250\,\mathrm{m}^2$ with some units less than $100\,\mathrm{m}^2$) the investment yield, without a local authority pre-let, would be substantially above that used earlier in this chapter (9% leasehold for a long lease with upward only rent reviews) due to the greater risks and management problems mentioned above. But with a single tenant of undoubted covenant these problems are eliminated and the investment yield should be comparable or even below the investment yield used in earlier examples where the head leaseholder was paying a rent reviewable on an upward only basis. With a pre-let to a local authority the lease granted to the investor (or developer) will be at a peppercorn rent and it will be the local authority who will pay rent on an upward reviewable basis related to rental value rather than rental income. A yield of 8.5% has therefore been used in the following example.

Because development risks also are now reduced, due to the pre-let to the local authority, the developer's profit margin has been reduced to less than a 10% mark up, also reflecting that the pre-let rent is a guaranteed minimum figure. Many of these schemes are undertaken by builders rather than speculative developers and they take a building profit only, so that the investment and development yields are the same. The leaseback rent is determined by calculating the gearing ratio which the developer estimates will produce the required return on expected (but not necessarily actual) costs.

At the date of the tender the developer's calculations would be as follows:

Estimated income
    (as in previous example)   £300 000 p.a.
Estimated costs
    building costs, fees, finance, etc.
    (as in previous example)   £1 946 000
Development yield
    Initial leasehold investment
    yield of say 8.5% plus annual
    return of say 0.75% p.a. to give
    a total required development yield
    of say 9.25% p.a.   0.0925   £180 000 p.a.

Estimated top slice income (or profit rent)
    to local authority   £120 000 p.a.

$$\text{Gearing ratio required by developer} = \frac{180\,000}{300\,000} \times 100\% = 60\%$$

The leaseback rent that the developer would require would therefore be 60% of the eventual full rental value, when the scheme is completed, or £180 000 whichever is the higher. The following calculation shows what the developer's and local authority's actual income might be on completion of the scheme on three different assumptions

1. The local authority achieve 100% occupancy at a FRV 10% above the developer's estimates made at the tender date;
2. The local authority achieve 100% occupancy at a FRV 10% below the developer's estimates;
3. The local authority achieve a 90% occupancy level at the same FRV per unit as originally estimated by the developer.

Table 14.1 Lease and leaseback apraisal

	100% occupancy FRV +10%	100% occupancy FRV −10%	90% occupancy at FRV
FRV	£330000	£270000	£300000
Actual income	£330000	£270000	£270000
Developer's income @ 60% of FRV	£198000	£162000	£180000
[Developers guaranteed minimum income]	[£180000]	[£180000]	[£180000]
Therefore developers income	£198000 (+10%)	£180000 (+0%)	£180000 (+0%)
Therefore L.A.'s income (less interest on land acquisition costs)	£132000 (+10%)	£90000 (−25%)	£90000 (−25%)
Developer's profit (% mark up on costs)	19.7%	8.8%	8.8%

Providing the developer's costs do not increase above the estimated figure, the developer will achieve his required return or more. Control of costs is the risk the developer takes and is why a small mark up of maybe 10% will probably be required when determining the gearing ratio. However, as stated earlier, many of these schemes are developed by builders who will estimate their costs very accurately, particularly in view of the small size of these schemes which can therefore be built over a short timescale of months rather than years.

From the local authority's point of view its top slice income is heavily exposed where actual FRV is less than the developer estimated or where not all units are revenue producing. The effect of a small (10%) drop in income dramatically reduces the local authority's profit rent and, in this example, voids in excess of 40% would eliminate it altogether. Clearly the local authority must examine the developer's estimate of FRV very carefully because if this is inflated or is optimistic, what might superficially look an attractive offer may turn out very differently when subletting occurs. Similarly, the authority must give serious thought to the likely level of voids. It is almost inevitable that with schemes of this type there will be voids. In the example used here, the relatively high land value acted as a substantial buffer and the local authority's top slice income was not as exposed as in many such agreements where gearing in the range of 80:20 to 65.35 (in favour of the developer) is more common as Table 14.2 illustrates.

Table 14.2 reproduces in simplified form part of the results of a tender for

Table 14.2 Lease and leaseback – developers' offers

Developer	Floor area m²	Estimated FRV £	Leaseback rent £	L.A. income £	Rent review gearing	Term of leaseback yrs	Interest to be acquired	Premium £
A	2085	224000	224000	0	100%	25	Freehold	925000
B	2085	224000	168300	55700	75%	25	Freehold	365000
C	2415	260000	182000	78000	70%	30	Freehold or long leasehold	150000
D	1675	144000	—	21600	85% of actual income	—	125 year ground lease	150000
E	2300	—	—	—	—	—	Freehold	300000
F	2305	173600	178600	−5000	80%	25	Freehold	351000

a small unit business space scheme in London that was held in late 1983. As can be seen most of the offers were made on a leaseback basis similar to that described above but with added ingredients such as premiums payable at the start or on completion of development (partly to cover outstanding site acquisition). One offer (from Developer E) was for the outright purchase of the freehold and one offer (from Developer D) was on a side by side basis with no guaranteed minimum leaseback rent. The offer from Developer E involved a guaranteed leaseback rent above the estimated FRV and only when the FRV had risen substantially above the original estimate would the gearing at subsequent rent reviews be applicable. Bearing in mind possible voids, this offer would not be attractive to the local authority unless very significant rental growth occurred. As there were no specific floor space figures included in the development brief, and as the bids are not on identical bases, financial comparison between the offers was not easy and necessitated a Discounted Cash Flow Appraisal over a 30-year period with assumptions made about rental growth and void levels.

14.9 Direct development by local authorities

The previous section discussed the reluctance of the private sector to become involved in small unit developments without an overriding lease of completed space to a local authority. The reasons for this were more to do with management and covenant problems rather than rental levels and viability although not exclusively. During the economic recession of the early and mid-1980s in many midland and northern parts of the country demand from manufacturing industry was, not surprisingly, very weak and rental levels were low with little rental growth. Even with a local authority covenant the private sector was reluctant to build small industrial units even where IBAs were available. Local authorities and other public sector agencies had no alternative but to intervene and develop schemes without private sector developer involvement. In parts of the north some schemes, particularly managed work space developments, have been developed by local authorities with heavily subsidized rents as a means of encouraging new start-up firms to take space and create employment. In some schemes the rents charged on an inclusive basis hardly even cover the annual running costs, let alone provide a return, however poor, on capital development costs. These schemes important as part of economic development policies to stimulate local economies, are an extreme example of direct local authority involvement but there are numerous other examples of local authorities building more traditional small unit estates and earning a reasonable return on capital.

   When a local authority acts as developer there are two main problems it faces. One is the organization of the development process and particularly the construction process and the other is the finance required and the constraints imposed by central government on capital expenditure and borrowing. Both these points have been examined in some detail in Chapters 11 and 12. The two most common ways of funding direct development have been the finance lease and deferred purchase.

The disadvantage of the former was that as a lease was granted to a finance company, prescribed capital expenditure was incurred, but at least this was notional and did not count as borrowing. So long as capital receipts could be generated this was, and still is, a possible solution. The advantage is that the lessee (the finance house) can claim the capital allowances although these are less important now than they were in the early 1980s.

The normal arrangement would be for a 30 plus year lease of the site to be granted to the finance house which would then leaseback to the local authority which would normally be appointed as agent in the construction of the units. The leaseback rent to cover the repayment of the cost of development would be equivalent to a mortgage with a variable interest rate based on six months LIBOR. Depending on the local authority involved, the interest rate may be a small margin above LIBOR.

The other main alternative is deferred purchase. This involves an agreement with a finance house, which provides the capital for development and usually appoints the local authority as agent to carry out the work. The capital cost is then repaid in annual instalments, i.e. the capital cost is effectively spread, or deferred, over a number of years. Until 22 July 1986, the main attraction of deferred purchase agreements was that they did not count as borrowing or capital expenditure. Although capital from a finance house was used, there was no lease and leaseback and so no notional capital expenditure. The finance house had no security, unlike with a finance lease, other than the local authority's word. Annual repayments were expenditure but as the total cost was spread over many years the annual amount was relatively small. However, all this changed after the Minister's statement of 22 July 1986 and the total cost now counts as one off capital expenditure. But like the finance lease no borrowing is officially incurred! Repayment of capital and interest is similar to the finance lease and is usually spread over a minimum of 5–10 years which is usually extended to 20 plus years to reduce the annual capital repayments. Capital repayments, however, can be deferred and interest during development can be rolled up (until income is produced from occupational tenants). Interest rates are normally slightly higher than in finance leases as the funder does not receive any capital allowances since it has no legal interest.

The numerical example below illustrates these two approaches using the same scheme as previously. It has been assumed that a design and build contract has been used, or a developer on a project management fee basis. This has increased the development costs by say 5%. Repayments have been taken over 30 years at a rate of interest half of one per cent above 6 months LIBOR and include capital repayment. Two scenarios are included.

Table 14.3 Local authority development

	Successful outcome		Unsuccessful outcome	
Actual income		£330000 p.a.		£270000 p.a.
Actual costs				
development costs	£2141000		£2141000	
project mgt. @ say				
5% of build costs	£107000		£107000	
	£2247000		£2247000	
Annual repayments (interest and capital @ say 9.25%)	0.10	£225000 p.a.	0.10	£225000 p.a.
L.A. income (less interest on land aquisition costs		105000 p.a.		45000 p.a.
L.A. equity		100%		100%
Interest payments on land acquisition say		£105000 p.a.		£105000 p.a.
(See assumptions at beginning of this chapter)				
Therefore initial net income		£0		−£60000 p.a.

NB. L.A. receive 100% of the equity. If and when rental income increases, as repayments are fixed, 100% of all increases will be received by the local authority.

The first assumes a 5% increase in development costs but a 10% increase in rental income, the second also assumes a 5% increase in development costs, but a 10% reduction in rental income (due to voids). A direct comparison can therefore be made with the lease and leaseback scheme illustrated in the previous section.

Any comparison of the calculations in Table 14.3 with previous appraisals, particularly the lease and leaseback approach in the previous section should be done with caution. In many cases a comparison in practice would be theoretical because, if both approaches were possible, the lease and leaseback approach involving the private sector would be pursued as it is less risky for the local authority and involves fewer management and financial problems. Nevertheless, it should be emphasized that in return for taking all the risk the financial rewards could be greater, particularly as the funding arrangement leaves the local authority with 100% of the equity.

For those local authorities which have programme or partnership area status under the 1978 Inner Urban Areas Act there is an additional calculation which is worth explaining. Schemes approved by the Department of the Environment (DOE) receive a 75% grant (although this approval is unlikely to be forthcoming if private sector involvement is possible). However, there is also a DOE clawback which operates on any profits from development (i.e. 75% of any surplus income, once loan repayments have been made, must be paid back to the DOE). Clearly this arrangement is meant for socially worthwhile, but financially marginal, schemes. As a result the local authority is unlikely to make a high return due to the 75% clawback, but on the other hand, is unlikely to lose money due to the 75% grant. The following calculation illustrates this and is an extension of the above appraisal assuming that the site costs the local authority £1 050 000 (this was the assumption made in Section 14.2) and includes infrastructure costs and interest charges.

Clearly in this example even in the unsuccessful outcome the local authority receives an income and there would have to be a very high void level for this income to be eliminated. Nevertheless, it is worth emphasizing that such an arrangement would only receive DOE approval where the scheme was likely to be marginally profitable.

14.10 Conclusion

Industrial and business space developments occur in a large variety of forms and in a large variety of locations; from large industrial estates, business

Table 14.4 Local authority development – I.A.P. funded

		Successful outcome		Unsuccessful outcome
Actual income		£330 000 p.a.		£270 000 p.a.
Actual costs				
Site cost (including interest)	£1 050 000		£1 050 000	
Development costs (including project mgt. fee)	£2 247 000		£2 247 000	
Total costs	£3 297 000		£3 297 000	
25% paid by L.A.	£824 000		£824 000	
Annual repayments say	0.10	£82 500 p.a.	0.10	£82 500 p.a.
Net income		£247 500 p.a.		£187 500 p.a.
25% to L.A.		£62 000 p.a.		£47 000 p.a.

parks and one off high-tech business units at one extreme to small unit industrial developments and managed workshop schemes which may be unprofitable or only marginally profitable using normal private sector financial criteria. The method of implementation and the type of financial agreement will naturally vary depending on the type of scheme, its location and the interest shown by the private sector – horses for courses.

Three main types of financial appraisal have been considered in this chapter corresponding roughly to broad categories of development and the assumed risks from a private sector viewpoint. In reality there will be a myriad of different arrangements often tailor-made to individual schemes. This chapter has discussed and illustrated the most common types but is not meant to be, nor could it ever be, exhaustive.

References

Baum, A. and Mackmin, D. (1981) *The Income Approach to Property Valuation*, 2nd edn, Routledge and Kegan Paul, London.

Britton, W., Davies, K. and Johnson, A. (1989) *Modern Methods of Valuation*, Estates Gazette, London.

Byrne, P. and Cadman, D. (1984) *Risk, Uncertainty and Decision Making in Property Development*, E. & F.N. Spon, London.

Cadman, D. and Austin-Crowe, L. (1983) *Property Development*, 2nd edn, E. & F.N. Spon, London.

Darlow, (ed.) (1983) *Valuation and Investment Appraisal*, Estates Gazette, London.

Hertz, D. and Thomas. H. (1983) *Risk Analysis and its Applications*, John Wiley, New York.

Jolly, B (1979) *Developmental Properties and Techniques in Residual Valuation*. Property valuation handbook CALUS, CEM.

Jaffe, A. (1979) *Property Management in Real Estate Investment Decision Making*, Lexington, Mass.

Marriott, O. (1967) *The Property Boom*, Pan Piper, London.

Morley, S. (1988) *Valuation and Development Appraisal*, 2nd edn (ed. C. Darlow), Estates Gazette, London.

Pilcher, R. (1973) *Appraisal and Control of Project Costs*, McGraw-Hill, Maidenhead, Berks.

Ratcliffe, J. and Rapley, N. (1988) *Development properties*, in *Valuation – Principles into Practice*, 3rd edn (ed. W. Rees), Chap. 13 Estates Gazette, London.

Index

Advance payment schemes 219
Assistance for Exceptional
    Projects 140
Assisted Areas, *see* Development
    areas
Audit Commission   141

B1 Use Class 55, 67
Back to back 217, 219, 223
Baker, Kenneth 114
Bank base rate 10
Bank finance 38
Barlow Report (1940) 117
Barter arrangements 220
Bottom slice 179, 280, 281, 288, 289
Bow Group 147
Break even rent 260, 269
British Urban Development
    (BUD) 144, 190
Brittan, Leon 113, 116, 143
Brown, Lord George 122
Building costs 251
Business use space, definition 67
Business in the Community
    (BIC) 144

Cambridge Phenomenon, The 70
Capital fund 210, 216

Capital receipts 210, 213–17, 221–6,
    285
Car parking 65, 73
Cascade 214, 215
Cash flow 261–7
Cash limit 212
Channel Tunnel 114, 160
City Action Teams 144
City Grant 37, 123, 144, 145, 184,
    188
Cladding 64, 75
Clement (VO) v. Addis Ltd. 134
Commercial Paper, Sterling 45
Community Development
    Project 118
Community Land Act (1975) 161,
    209
Consumer price indices 11, 30, 36
Contingencies 252
Covenant schemes, *see* Deferred
    purchase argreements
Credit squeeze 18, 21

Deep Discount Stock 44
Deferred purchase agreements 187,
    212, 218, 219, 222, 225, 294
Demand assessment 241

Department of Economic
    Affairs 118
Department of the Environment 122
Department of Trade and Industry,
    'DTI – the department for
    enterprise' 140, 142
Derelict Land Act (1982) 123
Derelict Land Grant 123–4, 128, 144
Development areas 118, 119, 122,
    128, 137
Development brief 200
Development control 149
Development cost 88
Development Gains Tax 18
Development Land Tax 133, 160
Development team 203
Direct development by local
    authorities 179–82, 187–9,
    198, 199, 202–4, 293–6
Disposal notice 184
Dobry Report (1975) 150

Economic development
    companies 188, 220, 228, 232
Educational Priority Areas 118
Employment 12–6, 60
English Estates 90, 144, 176, 189
Enterprise agencies 105
Enterprise boards 103, 107, 185
Enterprise Zones
    Dudley EZ 130
Environment, working 73
Eqities 43–4
Equity share, see Gearing
European Economic Community
    (EEC)
    Regional Economic Development
    Fund 114, 118, 122, 137
External appearance 65, 75

Fees 33, 42, 251, 255, 256
Finance 253–5, 257, 262
Financial appraisals 112
Financial deregulation 114

Floor loading 64, 75
'Footloose' industries 55
Forecourt 65, 73
Four slice 274–87
Frame-building construction 63, 74
Freeports 122
Funding 90
    see also Finance

Gearing 39, 192, 275, 278–82, 287,
    290
General Development Order 157,
    158
Golden horn 114
Golden triangle 30, 114
Greater London Development
    Plan 167
Green belt 116
Ground rent 177, 192, 193, 277–87

Hall, Peter 132
Heating 64, 75
Heseltine, Michael 132
Housing and Planning Act
    (1986) 129, 152
Howe, Sir Geoffrey 132

Incubator (workspace) 79, 90
Industrial Building Allowance
    (IBA) 184–9, 217
Industrial Development Act (1982)
    Section 8 projects 140
Industrial Development Certificates
    (IDC) 18, 113, 117, 176, 178,
    183
Industrial land values 58
Industrial Transfer Act (1928) 139
Inflation 11, 30, 36
Inner area programme 135, 184,
    185, 212–4, 223, 296
Inner Urban Areas Act (1978) 119,
    122, 134
Inner cities
    Action for Cities 128–9, 142–3, 145

Innovation centre 80, 83, 92, 93, 95
Innovation grants 143
Institutional Lease 27, 77
Insurance companies 48
Internal rate of return (IRR) 265, 266
International Monetary Fund
    (IMF) 118
Investment risk 25
Investment yield, see Yield
Investors in industry 114

Jenkin, Patrick 152
Joint company, see Urban
    Development Agency

Key sector 209

Lamont, Norman 119
Land registers 141
Land values, see Industrial land
    values
Lands Tribunal 171
Law of Property Act (1925) 171
Layout 63, 73
Lease and leaseback 196, 197,
    211, 216–8, 225, 287, 296
Lease terms 77
Leasing finance 42, 77
Leverage 39
LIBOR (London InterBank Offered
    Rate) 41
Lifting the burden (1985) 119, 150
Lighting 65, 75
Liquidity 39
Loading door 64
Loan sanction 208, 209, 212
Local authority companies 107
Local Government (Social Need) Act
    (1969) 118, 134
Local Government Act (1972) 170
    s.123 197, 202, 230, 231
    s.137 107, 210, 216, 227, 228,
        231, 232

Local Government Planning and
    Land Act (1980) 122, 130, 141,
    184, 197, 212, 228–30
Locally Determined Sector (LDS)
    209–12
Location 25, 26
Low tech 57

Managed workshops
    workspace 79, 82, 83, 89, 90, 93
Market research 239–44
Mid tech 57
Mortgages 17, 45

National Audit Office 128, 132, 139
National Loans Fund 131
Net asset value (NAV) 43
New enterprise workshops 83, 89,
    93
North–south divide 112
Notional capital expenditure 213,
    217, 218
Notional capital receipt, see capital
    receipt
NPV 265, 266
Nursery units 57

Office Development Permits (ODP)
    18, 117
Oil prices 9
Organization for Economic
    Co-operation and
    Development (OECD) 9
Out of town offices 68
Owner occupation 50

Participation (clause) 280
Partnership 191, 193, 194, 199–201,
    273–93
    schemes 108
Pension funds 49
Phoenix initiative 190
Planning agreements 158, 169–72

Planning conditions 157–8, 168–9
    legality of 172
Planning gain 159
    examples 163–5
    tests of reasonableness 171–2
Platzky, Sir Leo 130
Powell, Enoch 118
Pre-let 179, 191, 289, 290
Prescribed expenditure 213, 215,
    218–22
Probability analysis 271, 272
Profit 256–61, 264, 265, 276–8, 284
Project management 203
Property Advisory Group 156, 165,
    168, 171
Property crash 18
Public sector borrowing
    requirement 141

Regional Development Grant 122,
    137–40, 142
Regional Economic Planning
    Councils 118, 122
Regional economics
    regional imbalances 113, 116, 117
    regional incentives 116, 117
Regional Selective
    Assistance 137–40, 142
Rent cover 258
Rental growth 27, 61, 69
Rental value 56
    rack rent 249, 278, 279, 290
Residual valuation 244–9, 261
Retail Price Index 11, 30, 36
Return, see Profit
Return on capital 39
Ridley, Nicholas 133
Risk 256, 257, 277, 278
Roof 64, 74
Royal Opera House 172

Sale and Leaseback 17, 47
SANE 147
Scenario 269–72

Science Research Parks,
    definition 69
Section 52 agreements, see Planning
    agreements
Seedbed units, see Seedbed centres
Seedbed centres 57, 79, 83, 89, 93
Selection of developer 201, 202
Sensitivity analysis 258, 267–9
Serplan 115
Service charge 89
Services 65, 76, 79–84, 91–4, 105
Shaw, Sir Giles 143
Side by side 196, 274–87, 292, 293
Silicon Valley 70
Simplified Planning Zones 150,
    152–5
Simulation 272
Site analysis 243
Site cover 62, 73
Site value 257, 258
Special Areas Act (1934) 113, 117
Stamp duty 33
Structure plans 116
Superannuation fund 210
Supply assessment 242

Taxation 22, 33, 37, 42, 50
Tebbitt, Norman 119, 143
Technology centres 79, 83, 92
Tenure 66, 76
Third London Airport 160
Toilets 65, 76
Top slice 179, 280, 281, 288–91
Tory Reform Group 147
Town and Country Planning
    Acts 20, 55, 67
    Town Country Planning Act
    (1971) 228–30
Trade Parks, definition 71
Two slice 287–93

Unitization 46
Urban Development Agency
    190

Urban Development
    Corporations 37, 122, 130–2,
    144
  mini UDCs 131
  London Docklands Development
    Corporation 132
Urban Development Grant
    (UDG) 123, 128–9, 144, 184
Urban Programme 80, 83, 90, 118,
    122, 134–7, 143
  traditional 134, 135
  Urban Programme management
    initiative 135
  see also, Inner area programme

Urban Regeneration Grant
    (URG) 123, 129–30, 144, 184
Use Classes Order (1987) 55, 67,
    150, 156–9

Voids 192, 288–93

Walker, Peter 118
Widdicombe report 106, 107

Yield
    development 258, 277, 278, 284
    investment 249